The Care of Things

THE CARE OF THINGS

Ethics and Politics of Maintenance

Jérôme Denis
and
David Pontille

Translated by Andrew Brown

polity

First published in French as *Le soin des choses. Politique de la maintenance* © Editions La Découverte, Paris, 2022.

This English translation © Polity Press, 2025.

The translation of this book was supported by a grant from the Centre de Sociologie de l'Innovation, Mines Paris – PSL University.

Excerpt from 'Atlas' in *Selected Poems*, 2013, by U. A. Fanthorpe. London: Enitharmon Press. Reproduced by permission of Enitharmon Press, www.enitharmon.co.uk

Excerpt from *Ecologies of Knowledge: Work and Politics in Science and Technology*, 2013, ed. by Susan Leigh Star. Albany, NY: SUNY Press. Reproduced by permission of SUNY Press.

Polity Press
65 Bridge Street
Cambridge CB2 1UR, UK

Polity Press
111 River Street
Hoboken, NJ 07030, USA

All rights reserved. Except for the quotation of short passages for the purpose of criticism and review, no part of this publication may be reproduced, stored in a retrieval system or transmitted, in any form or by any means, electronic, mechanical, photocopying, recording or otherwise, without the prior permission of the publisher.

ISBN-13: 978-1-5095-6238-1 – hardback

A catalogue record for this book is available from the British Library.

Library of Congress Control Number: 2024940371

Typeset in 11.5 on 14 Adobe Garamond
by Fakenham Prepress Solutions, Fakenham, Norfolk NR21 8NL
Printed and bound in Great Britain by CPI Group (UK) Ltd, Croydon

The publisher has used its best endeavours to ensure that the URLs for external websites referred to in this book are correct and active at the time of going to press. However, the publisher has no responsibility for the websites and can make no guarantee that a site will remain live or that the content is or will remain appropriate.

Every effort has been made to trace all copyright holders, but if any have been overlooked the publisher will be pleased to include any necessary credits in any subsequent reprint or edition.

For further information on Polity, visit our website:
politybooks.com

Maintenance is a drag; it takes all the fucking time.
The mind boggles and chafes at the boredom.
Mierle Laderman Ukeles (1969)

Once something is perceived, the action of perception
continues indefinitely, changing and being changed
by other events near it, sometimes resonating,
sometimes clotting up or clumping up,
sometimes fading into background noise.
Susan Leigh Star (1995)

And maintenance is the sensible side of love,
Which knows what time and weather are doing
To my brickwork; insulates my faulty wiring;
Laughs at my dryrotten jokes; remembers
My need for gloss and grouting; which keeps
My suspect edifice upright in air,
As Atlas did the sky.
'Atlas', U. A. Fanthorpe

Some things last a long time.
Daniel Johnston

Contents

Acknowledgements	ix
Introduction	**1**
The art of making things last	3
A transfer of attention	7
The vocabulary of humans and things	10
Pathways	13
1. Maintaining	**16**
Beyond innovation	17
Repair and breakdown	21
A daily pulsation	26
Neither heroes nor heroines	34
Reassigning attention	37
2. Fragilities	**41**
Societies repopulated with objects	48
The diplomacy of alteration	55
Care and things	58
3. Attention	**62**
Displacements	66
Multisensoriality	71
Expertise	76
Vigilance	83
Attachments	86
4. Encounters	**91**
Recalcitrance	94
Disassembly	98

CONTENTS

Transformations	102
Worries	106
The dance of maintenance	110

5. Time	**114**
Prolongation	125
Permanence	129
Slowing down	147
Stubbornness	157
From time to thing	167

6. Tact	**169**
Adjustments	171
Surprises	179
Heritage diplomacies	184
Pathways inspired by environmental ethics	198
Ethics and the care of things	207

7. Conflicts	**211**
Shortening the life of goods	218
The values of duration	224
The emancipation of use	227
Redistributed knowledge	233
The people of things	236
Responsibilities	238

| **Conclusion** | **242** |

| *Notes* | 250 |
| *Index* | 283 |

Acknowledgements

Linked in many ways to the mundane fabric of encounters which forms our existences, this work owes so much to the particular attention of those who, through their generosity and their curiosity, have helped us to nuance our arguments, sharpen our descriptions, and reformulate our proposals. We wish to express to them here our sincere gratitude for the care with which they accompanied the slow fermentation of the different materials assembled in this book.

From the initial intuition to the different occasions of writing, there were of course many stimuli, spread over a long period, but one initial spark was decisive. So we warmly thank Mathieu Potte-Bonneville for being able to hear in the jumble of our first ideas the echoes of a possible book and for having accompanied us with enthusiasm and kindness throughout this journey.

Whether they have read the intermediate versions of certain chapters or an advanced version of the manuscript, we are greatly indebted to the attentive readings, meticulous comments, and always very astute suggestions made by Cornelia Hummel, Nicolas Nova, Marie Alauzen and Astrid Castres.

Informal exchanges during meals, in the corridors, or during remote meetings which punctuated the weeks of lockdown up to the workshop dedicated to the first complete version of the text, this manuscript also developed within a formidable working collective, the Centre for the Sociology of Innovation, where for so many years an art of precise discussion and attentive criticism has been cultivated. For their encouraging support and always relevant suggestions, we are extremely grateful to Madeleine Akrich, Victoria Brun, Béatrice Cointe, Jean Danielou, Liliana Doganova, Quentin Dufour, Daniel Florentin, Clément Gasull, Cornelius Heimstädt, Brice Laurent, Catherine Lucas, Alexandre Mallard, Kewan Mertens, Morgan Meyer, Fabian Muniesa, Mathilde Pellizzari, Florence Paterson, Vololona Rabeharisoa, Loïc

Riom, Roman Solé-Pomies, Didier Torny, Frédéric Vergnaud and Alexandre Violle.

At one-off or more regular meetings, in seminars or when texts intended for academic journals were reviewed, we benefited from the wise reactions and valuable advice of Antoine Hennion and Francis Chateauraynaud. The embryonic and partial versions of our arguments benefited greatly from the attentive listening of those two demanding readers, who hold a special place in the landscape of French pragmatist sociology.

The small community that has formed in recent years around maintenance and repair studies has constituted a thriving and refreshing environment, where our first investigations and collective reflections matured in an open atmosphere. First of all, a big thank you to Tomás Sánchez Criado and Blanca Callén in memory of our very first playful discussions of maintenance and care. We also remember the many rich conversations that sprang from the initiatives launched by Ignaz Strebel, Alain Bovet and Philippe Sormani in Zurich in 2014, then in Rome in 2015; Lara Houston, Steven J. Jackson and Daniela Rosner in Denver in 2015; and Andrew Russell and Lee Vinsel in Hoboken in 2017. During these meetings, Steven J. Jackson and Christopher Henke became regular interlocutors who shared with us their advice, and sometimes their perplexity, in the face of our investigations and our reflections, all in the interests of fruitful reciprocity. We would also like to thank, for the richness of their contributions, the participants in the various sessions that we organized between 2016 and 2022 at the conferences of the European Association for the Study of Science and Technology and the Society for Social Studies of Science. We also have fond memories of the journey made with Alexei Yurchak during his stay in Paris in 2018. Finally, a special greeting to Fernando Domínguez Rubio, who sent us a mysterious signal from San Diego in the spring of 2015, assuring us that his *Mona Lisa* and our metro signs had a great deal in common. Since then, the fascinating conversations and sudden bursts of laughter have never stopped.

The production of a book always extends beyond official work spaces to encroach in multiple ways on family time and friendly relations. Our thanks go to Michel and Annie Delgoulet, who offered us a haven of peace in Corrèze for a few days, sheltered from the upheavals of the pandemic: this place proved particularly conducive to writing. Finally, we are immensely indebted to Catherine and Alexandra, who helped us daily to make a place for this developing manuscript in the domestic and professional economies of our respective homes.

Raphaël, Alix, Nathan and Zéphyr: this book, which you will perhaps read one day, is also intended for you. We hope that it will stimulate your sensibility as much as your presence in the world nourishes ours.

Introduction

6:30 a.m. The alarm goes off. I get out of bed and have a slow stretch – a habit that I adopted after some particularly severe lower back pain. In the bathroom, my still sleepy ears are annoyed by the repetitive sound of the tap which has started dripping again. But I know that as a vintage model it doesn't just have aesthetic advantages: it has a gland housing, so all it should need is a quick turn of the spanner. I'll get down to it this evening, or during the weekend. Mustn't forget the silicone seals, either, I think wearily as I step into the shower, my eyes once again drawn to the black stains already accumulating on the smooth surface. If I leave them like this for too long, I run the risk of water infiltrating into the neighbours' flat downstairs.

7:15 a.m. I'm washed and dressed and have a first sip of my coffee, but it tastes odd. After a brief hesitation, I grab the bottle of white vinegar that I keep with the other household products. Having set aside the rest of the coffee in a thermos I programme the coffee machine to do a vinegar cycle, which shouldn't take too long. Anyway, I'm staying at home this morning. A technician sent by the condominium landlord is coming round to check our radiators, which seem to be having problems. Since the second Covid lockdown, I've had a high-speed internet connection at home. Many meetings are still being held remotely, and everything works perfectly, at least on a technical level.

11:00 a.m. The technician has arrived. I watch him place his hand on different parts of one of the radiators. He seems to be listening for sounds coming from the inlet pipe. He reminds me about the need to bleed the radiators regularly, to prevent air stopping hot water from circulating properly. After several minutes circulating from one room to another, he tells me that he has doubts about the expansion tank, a part of the collective boiler which is getting old. He suggests doing a test. He'll pump up the tank, increase the pressure again and see how it behaves. If, after a few days, the pressure has dropped, the part will need to be replaced. He'll check with the landlord.

12:00 p.m. I head to the train station to catch a train to Paris, where my office is situated. In the street, I pass a construction site which wasn't there yesterday. The technicians of the electricity distribution company have dug up the road and are inspecting a tangle of pipes and cables. I don't have time to stop, but can't help but glance at the gaping hole. I'm both shocked and impressed to see the state of these networks which supply the entire neighbourhood – the bakery, the apartment buildings and the clinical laboratory. String, adhesive tape and 'splints' made of unidentified materials bear witness to frequent repairs, over a long period of time, most of them undoubtedly routine even though some must have proved crucial. As I exchange a look with one of the men in hi-vis jackets emerging from the breach, I wonder: how can all this still work on a daily basis?

12:20 p.m. From the window of the commuter train, I spot the same hi-vis jackets being worn by a group busy checking the tracks of a temporarily closed railway line. Looking at them, I think of the terrible accident which took place in Brétigny-sur-Orge in July 2013. The presence of these workers, to whom no one seems to be paying attention, reassures me. Without them, who knows how many derailments would have already occurred?

3:00 p.m. Two people come into my office and stare at the ceiling. After greeting me, they use a pole to push a sort of glass cover onto a fire alarm that I'd never noticed before. A puff of smoke comes out of the instrument. Nothing happens. One of the two inspectors writes something on a form, then both leave without a word and knock on the door of the neighbouring office.

6:30 p.m. It's time to leave; I need to collect my electric bike from the repair shop where I left it the day before for its annual servicing. At the crossroads, I take a look at the church which has been under restoration for several months. Five or six people are gathered in front. Two are in their overalls, the others in formal suits. They seem to be debating: their attention is sometimes turned towards the voluminous documents they are consulting, sometimes towards the facade of the edifice, which several of them touch with the palms of their hands as if they were trying to ask its opinion. The building has stood there for decades, but today its fate seems to be in the hands of these experts.

7:00 p.m. At the repair shop, I'm told there weren't any particular issues with my bike: they've changed the brake pads, tightened the cables and also the nut for the stand, which has an unfortunate tendency to come loose over time and vibrate, making a horrible racket. I'll also need to be careful with the battery. The electronic diagnostic gauge shows that its charging cycle isn't all that efficient. I should avoid discharging it too often. I take note of this, remembering in passing that a few years ago, I was given the opposite advice about my laptop – I should systematically empty the battery before recharging it. I mention the possibility of changing the bike rack: one of the struts has cracked, probably a result of the cobbled streets that are part of my usual route. Don't change it just yet, I'm told. The pandemic has vastly increased demand while disrupting supply chains. Times are tough for spare parts.

7:45 p.m. On the way back, I pass an Apple store. I can't help but think about the screen of my iPhone: its corner got cracked a few weeks ago. So far, the electrician's tape that I skilfully applied to stop things from getting worse has done the job. Apart from that, the phone works perfectly, so there's no hurry – not to mention how much it would cost if I took my phone in here. I could go to the little shop right next to the station, of course, but I hesitate. I'm not sure where they get their spare parts from! I'd rather get an authentic iPhone screen and do it myself.

8:00 p.m. I return home with just one idea in my mind: tightening the bathroom tap. But on a chair in the living room, I discover one of my children's favourite sweaters, badly threadbare at the elbow. A little note accompanies it, written with what I imagine is some urgency: can anything be done?

The art of making things last

This mundane day, recounted as in a diary or field notebook, is only partially fictitious. Certainly, the 'I' composing it melds together the two present authors, and the scenes collected here were not concentrated in a handful of hours. That said, they all come from our recent experience. None was invented, nothing was exaggerated. This day could well have existed. Which means it is, obviously, 'situated'. It tells the daily life of a man who works in an office in Paris and can afford to wait for a technician to arrive in the morning. He lives in an apartment in a suburb accessible by train and cycles. If we followed the journey of a female executive in a large company, a logistics worker, a stay-at-home mother or a service worker, the scenes would change radically. Likewise, they would be very different if we located ourselves near a block of low-rent apartments in a so-called 'deprived' neighbourhood in another Paris suburb, or if we moved to a village in, say, Burgundy. Obviously, the contrast would be even greater if this day's events took place in India, Cuba or Mali. Yet, in all these places, and with all these people, we could find moments that would highlight the same obvious fact that our series of sketches reveals, or rather reminds the reader of: human beings are not limited to living among a multitude of objects of various sizes and functions; they don't just use these objects, they also *take care* of them. Bicycles, boilers, churches, toilets, electrical infrastructures, sensors, coffee makers, railway lines, kitchen taps, telephones and clothes are not just items at the disposal of the women and men who use them. They are also things in which it is necessary to intervene, more or less regularly, more or less drastically, so that they can continue to exist. But why is this? Because most of the objects that make up the human world don't last. They get damaged, decompose and wear out in various ways and for different reasons. So they need to be kept in good repair, otherwise they will disappear. It is to this somewhat humdrum activity that this book is devoted: maintenance. It covers a set of actions with blurred boundaries, made up of very varied tasks, gestures, types of know-how and even theories, but they all share the same motive, the same concern. Maintenance is the art of making things last.

If we feel it is important to investigate maintenance, this is mainly because people rarely talk about it, or do so in too restrictive terms. Alongside other notions that occupy the media space – innovation, of course, but also repair and resilience – maintenance still seems largely neglected. It's as if it were too banal or too unproductive. At a time when 'making a difference' seems to be the only thing that matters, making sure that everything stays more or less in place and that countless objects simply continue their bland existence isn't a very alluring activity. And there's another problem: maintenance often concerns things which themselves are not very glamorous, or which are also largely absent from contemporary discussions. Thus, while numerous studies, from different standpoints, agree that the Western, or Westernized, world is going through a 'crisis of sensibility',[1] this seems only to concern the living world, from animals to mushrooms, from trees to viruses. Of course, we concur with this view, which reveals, often very eloquently, how negligence, and also more simply the absence of descriptive skills, lead to sterility in political debate and to forms of collective action being taken for granted which deserve to be discussed anew in the current environmental crisis. It is, however, striking how blind these appeals are to an important part of what constitutes the material fabric of the human world. It's as if we first had to get rid of all the objects among which we live in order to make ourselves truly sensitive to the immense multiplicity of beings who coexist more or less easily on this Earth. Or, more simply, it's as if the question of *sensitivity* to this part of the world was already settled – as if we already knew how to describe the relationships that humans have with these things, so that it was no longer worth dwelling on the question.

It is probably Baptiste Morizot – whose work, which we admire, we will be drawing on here on several occasions – who most clearly illustrates this position in an awkward scene in the introduction to *Ways of Being Alive*.[2] He recounts a wonderful time spent observing aerial life above the Col de la Bataille, a mountain pass in southeastern France. But after describing the marvellous ballet of the numerous species of birds which meet there before beginning their migration towards the south, he draws a curious comparison. To underline the inestimable value of what he and his partner have observed, he presents in unfavourable terms another spectacle, which he openly denigrates: that of a vintage car rally. What is he criticizing them for? He reproaches them for not paying attention to what's happening just a few metres above their heads. And precisely, by not paying attention again and again, explains Morizot, we come to cease taking certain beings into account. We stop trying to create a common world with them. In fact, he continues, while we already know how to give a good account of our relationships with technical objects, we don't (as yet) know how to speak correctly about our relationships with other living beings. We don't know how to make ourselves sensitive to the richness of these 'prodigious because *different*' entities.

But are we really capable of accurately describing the relationships between humans and things? What exactly do we know about what it really means, for example, to maintain an old car? Even more, what do we know about how things evolve, grow old and persist? And are we really capable of understanding the attachment of those who work to make them last, or even the obligations that bind humans to objects? Are we able to *pay attention* to everything that is at stake in this relationship – one that it is tempting to condemn for several reasons?

Of course, cars pollute; of course, an innumerable number of objects that make up our world contribute to damaging our relationships with other living beings. And there is no doubt, as Emmanuel Bonnet, Diego Landivar and Alexandre Monnin explain, that one of the crucial questions posed by the environmental crisis is that of the ways in which human collectives, at different levels, detach themselves from several of these objects and work towards the 'closure' of technical infrastructures that damage living environments.[3] But, as these same authors point out, this immense technical and democratic problem presupposes precisely that we stop relegating the heterogeneous set of artefacts that make up our world to the background of contemporary debates. Why should we imagine that attention to the living world should necessarily involve dismissing things and those who are attached to them? And how can we divide the world in these terms, in the first place? Can we really separate the people who take care of the objects with which humans continue to coexist from those who take care of the living world, as if the two were not compatible? Is the relationship with the living world itself so immediate that it can do without measuring instruments, tools to equip the senses, monitors, and of course vehicles to move around, as well as clothing suitable for exploring the countryside, venturing into the forest or hiking in the mountains? Who takes care of these things? Who ensures the correct settings? Who repairs them? Who maintains them? Is it really reasonable to ignore this part of the story?

On the contrary, we feel that, in addition to calls to develop our sensitivity to living beings other than humans, in partnership with them, we have everything to gain from working on our sensitivity to *things* – including, precisely, the lack of interest in maintenance, which shows how limited this sensitivity still is. This is the first political move that we wish to defend in the present work: we seek to foster a return to what Bruno Latour called the 'missing mass' of the social,[4] the multitude of objects with which, and thanks to which, women and men form a society. This return does not claim, obviously, that everything is inevitably good or just in this mass. On the other hand, it aims to remind us of that trivial anthropological invariant which states that humans are never entirely naked on Earth – that they inhabit the world alongside things. We assume that, by focusing on maintenance, we can cultivate a new perspective on the relationships humans have with those things that are always more – or less – than symbolic totems or functional tools. Maintenance, and those who practise it, can help us rediscover this missing mass, which seems to have gradually faded from the concerns of many thinkers, including Latour himself,[5] except as seen through the prism of the negative effects that those material entities have on the world.[6]

Why? This is the challenge of the present book: we assume that maintenance itself has a political significance. In many situations, the art of making things last is part of a form of relationship with objects that contrasts with what is usually foregrounded – not only when the supposed benefits of 'technical progress' are praised, but also when the materialist excesses of consumer society are criticized. Maintenance often means resisting obsolescence and breaking, for a time, the cycle of incessant replacement. But it also disturbs the principles of a version of 'circular economy' which only has eyes for production, consumption and recycling. On another level, maintenance also involves disrupting the projections of a desirable or worrying future, a future that grabs all our collective attention, sometimes to the point of paralysis. It means acting within the ordinary fabric of everyday life, here and now, without our concerns being fixated on the blinding horizon of an insurmountable crisis, yet to come.

It is the discovery of these politics of maintenance, barely outlined here, on which we wish to focus, by making ourselves sensitive to things and to those who take care of them.

A transfer of attention

This is no simple task. Firstly, a large number of maintenance activities still lie under the radar of most people's experience, just as they remain in the margins of the main forms of contemporary master narratives. Maintenance is a quintessential background activity that, very often, doesn't seem to matter. What is more, the concrete activities that comprise it are not very talkative. Interviewing the women and men who carry them out, or simply listening to them talk, is not enough to understand what they do. Maintenance involves a recurring contact with matter, a contact that must be directly observed if it is to be understood. Not because this means that we thereby access a greater truth than by listening to people, of course, but because within the ecology of the situations – in people's gestures, looks, words and attitudes – a certain relationship to objects unfolds. Taking full account of these relationships is an essential resource when it comes to being sensitive to the fragility of things and the different ways of taking care of them.

Maintenance draws on an attentional gesture; it is 'a way of looking at the world'.[7] It is this gesture that we wish to explore. But not just as one object of research among others – something which, having been placed at a necessary distance, remains completely external to us. The objective of this book is to develop our own sensitivity to this very gesture, learning from those who cultivate it. In this sense, we will seek to engage in what Yves Citton calls a 'meta-attentional engagement': to deploy an instrument thanks to which attention is 'plugged into the attentional experience of another more or less strongly subjectivized perception of the world, through which a certain reality is revisited'.[8] We want to make ourselves, and you, attentive to the attention of those who practise maintenance.

To do this, we will draw on several types of resources. They all take the form of stories. Some are well known, others less so. Some are part of the 'great' history of civilizations, others are mere anecdotes. All of them will serve as guides to better understand how things come to last, and how certain women and men contribute to their becoming. Among these stories are, first and foremost, our own investigations. We will be immersing ourselves in the eventful life of metro signs and in the daily activities of the maintenance workers who look after them within the Régie Autonome des Transports Parisiens (or RATP, the Paris public transport company).[9] We will also follow the graffiti removers who work every day to maintain the graphic order of the city of Paris by removing from the facades, as much as possible, those inscriptions that are deemed undesirable.[10] We will also discuss certain aspects of the asset management of water networks in France, to which Daniel Florentin devoted an investigation together with one of the authors of the present book.[11]

We will also rely largely on the cases presented in the work of our colleagues. In recent years, in fact, we have been able to discover a growing number of articles and books by authors from very varied disciplines (sociology, history, geography, media studies, computer science, philosophy, architecture, anthropology, etc.) which focus on maintenance activities. Through meetings, international conferences and more informal conversations, a small community has taken shape, eventually forming the outlines of a particular domain: maintenance and repair studies.[12] This book will be an opportunity to discover the richness of these studies and the variety of maintenance situations that they describe. Curious readers will thus be able to collect, via our citations and endnotes, a set of valuable references which implicitly form a provisional 'state of the art' for this field of research, which is still a hive of activity as we write these lines.[13]

Finally, we will also draw some examples from the press. These cases do not have quite the same descriptive density as those resulting from research. However, as we will see, they are not without either flavour or relevance. They will also remind us that, if maintenance is generally neglected, it can sometimes temporarily become highly newsworthy. It then breaks down into different public problems: even the very formulation of these problems proves to be highly instructive.

Telling stories, rather than pretending to present a global and well-ordered view of maintenance *in general*, is not a trivial matter.[14] In writing this book, we want to think *with* maintenance and, even more, *with* those who practise it. To do so, we cannot completely detach ourselves from the great variety of situations. We cannot abstract ourselves from the singularity of the material encounters which maintenance involves. We must observe, listen to and accompany those who take care of things. We need to assemble stories that allow their voices to be heard, without ours drowning them out (let alone replacing them) – stories that allow us to develop a vocabulary capable of heightening our capacity for astonishment and sharpening our sensibility.[15] Following the valuable advice of Dorothy Smith, we will seek less to act as spokespersons for the women and men who practise maintenance than to create, along with others, the means of amplifying their words and their actions.[16]

The vocabulary of humans and things

In order to successfully carry out this project and to (re)discover maintenance, we must be attentive to the vocabulary we use. In particular, we need to find the right words to describe the two main protagonists of our stories: humans, and the things they care for.

Certain maintenance activities are mainly carried out by women. Others, on the contrary, are largely reserved for men. Elsewhere still, the division of tasks is more complex. Although it is important to show these differences, we will not be able to adopt a standardized lexicon that can fully cover them. On the other hand, we will try to follow a simple rule by regularly marking the presence of both genders, in particular when habit or laziness of the imagination could lead one to believe in the sole presence of men in the situations described. At the same time, we will emphasize the role of women when their presence is more important. This is the general principle which we will try to respect while preserving the fluidity of the text as much as possible.

We will also sometimes mention 'humans'. In this case, the aim will be to mark a distinction between all women and men on one side, and the things that are being maintained on the other. This manoeuvre may seem a little incongruous at this stage, but we will see that it matters in certain cases, in particular because the relational nature of maintenance raises the question of the role that humans 'in general' allow themselves to play in the life of things.

Let's move on to things, now. To things or to objects? The question may seem artificial: we could easily argue in favour of a relaxed use of synonymy and assume that we can move from one term to another without worrying about the specific resonances of each. But this would mean depriving us of the subtleties that the two words convey and missing a fundamental aspect of maintenance. So we will not be using the two terms in quite the same way. Without adopting too rigid a protocol here, we will try to play on their difference so as to distinguish between, on the one hand, what is obvious and 'already there', which is treated as *objects*, and all that can't be taken for granted, all that can disappear, dissolve and crumble, and needs to be constantly actualized: *things*. In other words, objects are everything that can be used without being questioned, in an almost transparent way. Nobody is worried

INTRODUCTION

about them. They fall under what Bruno Latour, who proposes a similar distinction, calls 'matters of fact'.[17] What things are and become, on the other hand, lies at the heart of many different worries. They are 'matters of concern'. Things need to be taken care of if they are to continue to exist.

Clearly, this distinction does not relate to specific entities that could easily be divided up into the two categories, once correctly identified. Rather, the two terms refer to specific relationships, relationships in which these entities are caught up. Let's take an example. You treat your tap as an *object* when you simply use it to run (or not run) water in the bathroom washbasin. On the other hand, if you regularly listen to the sound of the drops that it can let fall, if you try to feel the degree of resistance it presents when being turned on and off, if you have an idea of the different elements that compose it and are familiar, even superficially, with their behaviour (if you know, for example, about the possibility of wear and tear to a ceramic cartridge left in the same position for too long, or the pressure that a damaged rubber seal is likely to exert), you are treating this 'same' tap as a *thing*. To put it more simply: maintenance amounts to apprehending as a thing something that, for others, or for the same person but on other occasions, is an object.

This choice of vocabulary obviously echoes the philosophy of Martin Heidegger, who himself distinguishes between the object and the thing.[18] However, he has a different political and moral objective. As Latour explains,[19] Heidegger a priori separates objects, which are the result of modern technology (inevitably cheap), from things, whose artisanal, even poetic, character ensures rich forms of connection to which we ought to return. Just like Latour, the present authors refuse to decide on the issue in advance: we wish to leave open the possibility for apparently banal and standardized objects to be treated as things. It is even the great strength of maintenance to make us rediscover, quite literally, artefacts that are generally undervalued.

Our suggestion, however, is closer to the famous distinction that Heidegger establishes between the 'readiness-to-hand' of objects of which we make ordinary use and the 'presence-to-hand' that charac- terizes those that break and stop functioning properly. However, we will see a little later that our proposition does not totally stick to Heidegger's, in particular because maintenance and its concerns do not need breakage in order to be carried out. Taking care of things is not conditional on their breaking down – quite the contrary.

Finally, rather than in Heidegger, it is in pragmatist philosophy that we must seek the source of our proposed distinction (a flexible one, as we have said) between objects and things, particularly in what William James calls 'pragmata': those 'things in their plurality' that are not just given.[20] Things are unfinished. They are constantly in the making. They are even still 'to be made', to use Étienne Souriau's eloquent term.[21] The women and men who take care of them are well aware of this, since they worry about them and participate directly in this fragile and uncertain 'making'.

Referring to *things* therefore allows us to underline the dimension, at once unstable and unpredictable, of the material entities apprehended by maintenance, including when they are massive and at first glance largely stabilized and inert. But that's not all. As the example of the tap suggests, the 'worrying becoming' of things in maintenance gives way to a certain density, a swarming or a certain thickness, which stands out from the transparency and apparent flatness of the objects. With things, materials start to matter. Numerous, heterogeneous, sometimes elusive, they form the moving reality with which we must deal to make things exist and last. As Tim Ingold reminds us, objects generally present themselves to us in their finitude, crystallized in a form that we very easily take for granted, without having any idea of the material actions that constantly pass through them:

> It is as though our material involvement begins only when the stucco has already hardened on the house front or the ink already dried on the page. We see the building and not the plaster of its walls, the words and not the ink with which they were written. In reality, of course, the materials are still there and continue to mingle and react as they have always done, forever threatening with dissolution or even 'dematerialization' the things they comprise.[22]

Maintenance presents itself as a sort of antidote to the blindness that Ingold regrets. It cannot deal with crystallized objects. In its apprehension of things, it focuses attention on the diversity of materials present and on their behaviour. Because the people who take care of things act straight from the flows of matter, they help us to become sensitive to those minute movements of materials, or 'intra-activity' as Karen Barad calls it,[23] which constitutes the fragile life of things.

INTRODUCTION

Pathways

This book is organized around individual words that designate the main questions and principal problems which an investigation of maintenance activities invites us to reconsider. We begin, in chapter 2, with the question of *fragilities*. We have already mentioned this point: by following people who take care of things, observing their actions and taking their concerns seriously, we find that a large number of objects, whose physical properties are generally taken for granted, are not as solid as one might imagine. From this point of view, maintenance de-centres (or re-centres) the gaze. By highlighting material fragility, maintenance sensitizes us to the modulations and deterioration that affect the material fabric of human societies. It entails a striking discordance, not only with the ordinary experience of many people, but also with the way in which numerous studies in the humanities and social sciences have tackled 'materiality' in their analyses by insisting exclusively on solidity and durability. Rather than denying or neglecting the wear and tear of things, maintenance starts from it. It sees it as a common condition, which forces humans to imagine different forms of diplomacy with matter. It is in this sense that maintenance can be seen as a way of taking care of things.

This fragility, however, is not obvious. That, indeed, is the problem. In chapter 3, we will look at the *attention* cultivated by those who take care of things in order to become sensitive to them, particularly in occupational situations. Maintenance involves contact with matter, and through this contact – which mobilizes the gaze, but also touch, hearing and smell – the people who take care of things strive to let these things express themselves. This uncertain inquiry, open to the unexpected, shows that maintenance is also an art of getting to know things on their very surface.

In chapter 4, we will continue to explore these *encounters*. Far from always remaining peaceful and delicate, the material interactions that punctuate maintenance may also generate tensions. Sometimes things resist care – especially when care can involve hand-to-hand tussles punctuated by sometimes quite spectacular disassembly operations. Far from sticking to subtle acts intended to preserve at all costs every aspect of the object concerned, maintenance turns out to be much more transformative than one might at first imagine. And encounters are often accompanied by a certain ontological concern: despite maintenance – because of it, in certain circumstances – the thing sometimes risks disappearing. To prevent this from happening, maintainers must give it room for manoeuvre and allow it to manifest itself: they must engage with it in a dance that humans are not the only ones to lead.

We will then return to the idea that maintenance consists of making things last, by exploring the types of *time* that are at work in this seemingly innocuous expression. Rather than simply constructing a purely 'social' time, maintenance is above all a way of making time a problem. By following this path, we will be led in chapter 5 to identify important differences in the problematization of duration. If, in certain cases, taking care of things simply consists of prolonging their use for an indefinite period of time, in other situations maintainers strive to produce the conditions for a quasi-eternity. Elsewhere, it is rather a matter of slowing down. Maintenance then amounts to fighting against the inevitable time of material deterioration, without ever claiming to stop it completely. Finally, things themselves sometimes persist in lasting well beyond what had been imagined by the humans who find themselves obliged to take care of them, more or less easily. These different ways of making time a problem are always closely associated with practical ways of organizing care. They also have a different ontological scope. Each form of maintenance goes hand in hand with a certain definition of the thing maintained and a more or less explicit identification of the elements that constitute it, elements on which care must concentrate in order to ward off the risk of seeing the thing disappear.

After exploring the problem of duration, we will question the other verb which makes up the expression 'to make things last'. In chapter 6, we will see that the action of 'making' is not obvious, and that the question of its scope, like that of its intensity, haunts the whole question of maintenance. By focusing on the problem of authenticity, we will see that taking care of things is a matter of *tact*. This is what the intersecting history of the debates over heritage conservation and over environmental preservation shows. The former was initially based, in its modern version, on the demand for strong action by men (more rarely women) on the life of things of value. The latter, on the contrary, was based on keeping humans at bay, setting them apart from the nature to be preserved. In recent years, the two fields seem to have come together in more nuanced attitudes, which give the dance of maintenance a much more subtle dimension.

Finally, in chapter 7, we will address a question that runs throughout the book: who takes care of things? The resistance of many groups around the world and the *conflicts* that have emerged around the 'right to repair' will help us understand that if maintenance is very often a 'dirty work' reserved for a poorly regarded segment of the population, it is also sometimes claimed as a vector of emancipation, a means of cultivating an alternative relationship with consumer goods, freed from the arbitrary constraints set in place by large industrial groups. These conflicts remind us of the existence of a 'people of things' who strive, sometimes in the less visible margins of contemporary capitalism, to take care of objects that they do not want to see disappear.

But before starting this journey,[24] we must return to the very term 'maintenance' and highlight the theoretical and political gesture which consists of placing it at the centre of our book. This is what we propose to do in the first chapter. For this, we will be drawing on Mierle Laderman Ukeles, a conceptual artist close to Marcel Duchamp and author of a manifesto for the art of maintenance in 1969.[25] She is a sort of godmother of maintenance and repair studies. Her artistic approach will help us better understand the movement, both perceptual and conceptual, which consists of bringing to the forefront of our descriptions activities that are generally ignored or insufficiently esteemed. By focusing on maintenance, a reproductive activity par excellence, one that is carried out in the ordinary run of daily practices, we can rid ourselves of the contemporary obsession with innovation. But because it operates within the fabric of continuity, maintenance differs more generally from the omnipresent idea of disruption. In this sense, it also allows us to distance ourselves from the vocabulary of repair and resilience, as these remain deeply marked by the language of crisis, accident and shock, all of which disrupt an initial order, a peaceful state that must be restored. With Ukeles, we will realize that maintenance operates in a completely different space-time. It is an activity anchored in the folds of the present, one that always needs to be repeated: a banal gesture which does not recognize any hero. That is precisely why it deserves our full attention.

1

Maintaining

On 20 July 1973, Mierle Laderman Ukeles was invited to the Wadworth Atheneum in Hartford, Connecticut, to stage her very first performance, entitled *Transfer: The Maintenance of the Art Object: Mummy Maintenance: With the Maintenance Man, the Maintenance Artist, and the Museum Conservator*. As the title suggests, Ukeles was not alone. Inaugurating a long series of collaborative works, she invited one of the institution's curators as well as one of its maintenance staff. At the centre of the performance was a work already on exhibition in the museum – an Egyptian mummy on loan from the Metropolitan Museum of Art – around which the three protagonists played their parts. Less than the mummy itself, it was its display case which was at the heart of the operation, with spectators being encouraged to pay attention to the special care it received. The performance was organized in three distinct moments. It began with the cleaning of the display case by the maintenance worker, who carried it out as he had done every day since the work had entered the museum. A maintenance task like any other – quite routine. Secondly, the worker's gestures were repeated identically by Ukeles, who had carefully observed them. This second phase ended with the affixing of a stamp to the display case, inscribing the expression 'Maintenance Art Work' in ink. In a gesture directly inspired by the work of Marcel Duchamp, the operation transformed the impeccably clean surfaces of the display case into proof that the work of cleaning was itself a work of art. The curator was then obliged to authenticate it and write a 'condition report' for the display case. Above all, he became the only person authorized to carry out future cleaning operations. Now handled by a curator, the actions carried out by the maintenance worker, and then by Ukeles, had thus changed their status. Maintenance had shifted into the realm of art.

The performance, a first opportunity for Ukeles to put to the test the principles of her manifesto for maintenance art,[1] had a considerable impact and marked a time of profound upheaval in conceptual art. What was she doing, exactly? How did she help to transform the gaze of the spectators, but also that of artistic institutions themselves? The implications of this 'transfer' are numerous, but one central point in Ukeles' work is particularly salient here, and deserves our attention. By making maintenance an art, Ukeles strives to highlight the 'reproductive work' that is usually left aside, remaining in the shadows of exhibitions, reserved for spaces and times from which the public is absent. Simultaneously, she seeks to unravel the obviousness of the creative gesture and to undermine the separation between the two forms of action, one highlighted in the name of originality, the other devalued because it is ordinary, a matter of mere continuity.

She has also explained on several occasions that her conceptual work was particularly inspired by her husband's experience in the urban planning department of New York City. In the official planning documents, she had in fact discovered that the management of the city, in particular its financing, was organized around a very clear differentiation between what fell under 'development' on the one hand, being valued and encouraged, and what related to maintenance on the other, generally reserved for professions that had little prestige in terms either of salary or of public recognition. This difference still largely structures the life of private and public institutions nowadays, right down to their accounting systems, which separate investment expenditure (strongly encouraged) from operating expenditure (subject to significant constraints). It was by identifying the parallel between this accountancy-based distinction and the obvious hierarchy at work in her own professional world that Ukeles decided to systematize the critique and practical deconstruction of the latter. How can we make art by moving away from the model of the individual inventor and his or her disruptive act? How can we make the actions of those who safeguard the material conditions of art a form of art in themselves, in a world where only what 'makes a difference' seems to matter?

Beyond innovation

Initiated in the early 1970s, this radical de-centring of the gaze is still relevant today. It seems even more urgent. In a world saturated with hymns to novelty and inspiration, where innovation is elevated to the level of a quasi-religion, claiming to have an interest in the multitude of interventions that simply aim to make things last is almost a militant act. In the United States, it is precisely in this respect that interdisciplinary reflection in and on the community of 'Maintainers' has been organized for several years. Its initiators, Andrew Russell and Lee Vinsel, two specialists in the history of computing, have published a series of columns in which they directly associate the need to worry about maintenance and maintainers with a questioning of the obsession with innovation and innovators, an obsession found both in the general press and in specialized publications and a large proportion of academic studies relating to different technologies. The title of their recent book on this issue – *The Innovation Delusion* – could not be more explicit: recognition of the importance of maintenance activities fuels a head-on critique of the political and economic positions that lead to innovation becoming the sole driving force of contemporary societies.[2]

At a time when the vocabulary of disruption has become essential in capitalist morality, directing our gaze towards maintenance thus consists of reversing the hierarchy of themes and foregrounding what constitutes continuity. As in art, where a vast series of maintenance activities makes possible the very existence of a creative gesture valued as individual and unique, the disruptive effects attributed to the geniuses of innovation could not see the light of day unless a considerable number of things remained unchanged, first and foremost the complex infrastructures that provide a reliable basis for mobility, telecommunications and the production and distribution of energy. The ability of these things to last, and the fact that everyone (including 'innovators') can rely on their stability, depends on countless operations dedicated to preventing other forms of undesirable disruption from occurring.

Refusing to establish creation and innovation as overarching moral values also changes the situation in terms of the type of objects in which we may be interested. By considering the cleaning activity of the museum maintenance worker, we realize, for example, that the display case, an essential element in the refined presentation of modern artistic institutions, plays an important role in the very existence of the mummy as exhibited to the public eye. It also becomes clear that the cleanliness of this display case, and therefore its transparency, must not be taken for granted.

The operation can be generalized. As soon as we take an interest in the thousand maintenance operations that punctuate our ordinary lives, a multitude of artefacts appear on the surface of the world. Just as display cases were not present in art history books in 1973, those ordinary objects are absent from the frescoes of modernity which describe technical progress and extol the merits of innovation by focusing on a very small sample of exemplary artefacts. Perhaps even more than the quantity of objects to which maintenance encourages us to pay attention, it is their condition that matters. In a regime of permanent innovation, the material and technological world, as described even in the daily press, is populated by artefacts with ever more complex functionalities, with their refined design and perfect appearance; but garages, repair workshops, car parks, cellars and gardens where one form or another of maintenance is carried out on a daily basis are occupied by banal and sometimes frankly rudimentary objects that are often obsolete in technical terms. Above all, these places are full of *worn-out* objects which bear the traces of what has sometimes been an already long life. If there is one essential thing that maintenance teaches us, or rather reminds us of, it is that we live surrounded by old trinkets which we continue to use most of the time without this causing any problem. The technical environment in which we operate does not have the brightness of the modernist pictures painted by those who, all year round, harp on the virtues of innovative technologies. Many of the objects composing this environment are old, their surfaces tarnished and scratched, not all functioning properly. And there's nothing dramatic about all this – quite the contrary.

It is probably the historian David Edgerton who best described the consequences of this repopulation of artefacts, notably on the occasion of the 2006 publication of his book *The Shock of the Old*.[3] The obsession with the idea of innovation is in fact much more than a bad habit adopted by suggestible journalists, or an abstruse stratagem repeated throughout PowerPoint presentations and posts on LinkedIn by self-proclaimed specialists. For years it has nourished the history of technology as such, most descriptions of which are guided by a sort of premium on novelty which marginalizes a myriad of objects whose material properties and uses are nevertheless crucial to understanding the contrasting evolution of the place of technology in the lives of human beings worldwide. Maintenance represents a considerable proportion of the uses that are rendered invisible by the almost exclusive focus on invention. It offers a privileged entry point for writing what Edgerton calls, with acid irony, a history of technology aimed at 'grown-ups of all genders' rather than 'boys of all ages' (generally white)[4] who are the audience for all those analyses which 'conflate technology with technological novelty'.[5] Less fantastical and more realistic, such a story strives to describe the technical objects in the situations it studies without a priori distinguishing between what is supposed to 'make a difference' in the lives of each person individually. Seemingly innocuous, this simple principle of symmetry is in fact very powerful and has significant repercussions. Because it takes into consideration our old trinkets, and studies in detail the long life of artefacts of all kinds, a life that is sometimes fraught with pitfalls and more or less subtle transformations, this history of technology brings about a major geopolitical shift. Its analysis can include not only practices usually ignored, but also a set of poor countries that had hitherto been ignored in narratives of modernity.

The huge work of continuity, and a myriad of worn-out objects that need to be taken care of: here are the main ingredients of the counternarrative that maintenance encourages us to write if we are to resist the blinding obsession with innovation. But is this really just what these two avenues lead to – a fight against the unhealthy relationship with technical progress and against the idea of creative genius? That's already a lot, of course. And there is no question of minimizing the significance of the struggle waged by those who refuse to submit to the dictates of disruptive innovation. But it seems to us that the critical significance of shifting attention to maintenance is even greater. In particular, it encourages us to question the foundations of another form of action which has been much talked about in recent years: repair. The terms maintenance and repair are frequently combined, and they can seem at first glance almost synonymous. However, on closer inspection, their uses do not quite describe the same operations. Above all, they do not refer to exactly the same problems. The notion of repair is closely linked to the ideas of breakdown and accident, and thus highlights very specific social, economic and political issues that only partially cover what maintenance sees as crucial. Without falling into an abstract nominalist debate, it is useful, still following Ukeles' conceptual path, to understand how choosing maintenance as a starting point for descriptions differs from concentrating on repair. This is all the more important as repair is a powerful notion, and lies at the basis of a large number of fascinating current investigations, some of whose hidden assumptions run the risk of partly reproducing the blindness generated by the obsession with innovation.

Repair and breakdown

It is impossible not to see in the growing success of the notion of repair a reaction to the crises that have been shaking the world since the beginning of the twenty-first century.[6] Nowadays, the term seems to be establishing itself as one of the main tools for criticizing the regime of permanent innovation and, more generally, the very idea of technological progress. From the acceleration of climate change to the collapse of biodiversity, the concerns it conveys resonate directly with the upheavals which are shaking what is now commonly called the Anthropocene – the era which sees the life of the entire planet as affected by human activity. Environmental crises, migratory crises, socio-economic crises, and finally the health crisis triggered by the appearance of SARS-CoV-2: all fuel the observation of a general breakdown of modernity. Faced with this diagnosis of a world that has become dysfunctional, barely viable, the idea of repair points towards an open horizon, while designating a moral imperative. If the world is broken through 'our' fault, it is up to us to repair it.[7] With repair, the outline of a form of saving action emerges – an assumption of responsibility that can adopt a very wide variety of concrete forms, from agriculture to welcoming migrants, including the regulation of finance capitalism.

In academic literature concerned with describing concrete situations (mainly in history, sociology and anthropology), this interest in repair very often goes hand in hand with an interest in maintenance activities, the two terms sometimes even being interchangeable. But while, in the vast majority of these studies, maintenance and repair are discussed simultaneously, it is the second term which is the focus of the most advanced conceptual activity. It is repair that must be described and understood above all, repair apprehended as an anthropological gesture, a truly human 'impulse'.[8] It is repair which is established both as an analytical tool and as a moral value. However, this notional primacy is not without consequences. In particular, it focuses attention on phenomena and activities whose relevant properties are not so foreign to the concerns of narratives centred on innovation.

Above all, as we have seen, the notion of repair is closely linked to the idea of a breakdown or accident. And its use is part of an already long history of the idea of dysfunction. The idea of breakdown has long been a valuable resource for all those who set out to deconstruct the grand narratives of technical progress. Very recently, two studies by authors close to our own concerns have started from the depiction of a catastrophe. Chris Henke and Ben Sims begin their book *Repairing Infrastructures* with a detailed evocation of the collapse, in Minneapolis on 1 August 2007, of the bridge which carries Interstate 35W across the Mississippi.[9] Russell and Vinsel, whose work we mentioned above, begin their analysis with a chapter entitled 'The problem with innovation', the first lines of which describe a series of explosions caused by a gas leak in the Canadian province of New Brunswick in 1986. By adopting this strategy, these authors are part of a line of work that uses this type of events in order to create the sense of a revelation. When technologies no longer work as they are supposed to and, even more, when their physical coherence disintegrates before our very eyes, it becomes possible to question their value without being burdened with a series of presuppositions about the linear nature of their trajectory or even of their political neutrality. Breakdown and breakage take on an epistemic value in these analyses. It is striking from this point of view to note the importance of the *Challenger* shuttle disaster in the 'technological turn' which took place within science and technology studies at the end of the 1980s and the beginning of the 1990s.[10]

The revelatory virtues of accidents and sudden dysfunctions are partly rooted in a reading of Heideggerian philosophy that could be described as 'optimistic'. In his description of the uses of technology, which we mentioned in the introduction, Heidegger draws a distinction between a technical object's two ways of being in the world:[11] 'readiness-to-hand' and 'presence-at-hand'. The first expression refers to the ordinary use of things. When wielded in a normal situation, a hammer – to take one of Heidegger's canonical examples – appears to the unobservant eyes of its user, and in his or her hands, a perfectly obvious tool, 'to hand'. It almost disappears, in favour of the transparency of its use and of the work it is helping to produce, the projected horizon of the action. On the other hand, if it is too heavy or poorly handled, the hammer sticks out and becomes problematic. It then reveals itself to be present 'at hand', all *too* present to the annoyed gaze of those who can no longer make obvious use of it. Its operation is no longer transparent and becomes the subject of a questioning. And the relationship with the world that the tranquillity of its use made possible is itself destabilized.

In science and technology studies, breakdowns and accidents are understood as formidable tools for problematization, in the good sense of a questioning. These events are supposed to shift the technologies of daily life from one form of being to another, from the obviousness of the transparent tool to an investigation of the enigmatic thing. Malfunctions are seen here as opportunities to bring to the fore aspects of technology that have so far been absent from the experience of its ordinary users.

In the different studies that draw on these opportunities, two main directions are followed, often in a successive and complementary manner. The first consists of taking the opportunity of the breakdown to discover the 'inside' of the technical object concerned. When it operates without any pitfalls, this object is in fact similar to a 'black box' that is unknowable to most people. By forcing the opening of the black box, the accident or breakdown reveals heterogeneous constituent elements previously 'folded' inside the object and rendered invisible by ordinary use. If we want to understand this effect of revelation and the resulting emergence of some of these components, we need simply remember the importance that Pitot probes and Stall alarms assumed for the French in the summer of 2009, following the crash of Air France flight 447 from Rio to Paris. More recently, the MCAS computer program, which Boeing engineers installed to give background assistance to 777 Max pilots by automatically correcting the plane's behaviour, suddenly found itself at the centre of public concern, and was the main subject of countless articles after two of these planes crashed within months of each other. Despite the horror that these tragedies represent, it must be recognized that they made it possible to bring to the attention of a very large number of people certain components essential to the operation of contemporary aircraft. These accidents, and the investigations that followed them, did much more than simply enrich the description of complex socio-technical assemblages. They brought these hitherto almost invisible components into politics by placing them at the heart of sometimes vehement debates, involving reparation processes (including criminal and financial) in which technical, economic, legal and even managerial considerations were closely intertwined.

The second revelatory aspect of a breakdown follows a diametrically opposite movement since it no longer consists of looking at what makes up the 'interior' of a technical object, but in pointing out what links it to an 'exterior' rendered intelligible by revealing the network of inter-dependencies in which the technology concerned is embedded. The most classic example in this field is undoubtedly the lessons that several researchers have drawn from blackouts in the United States – those moments of interruption in which electricity is no longer available, and the daily lives of the residents of large cities seem to be put on hold.[12] The electricity supply is the essential condition for a considerable number of activities and for the functioning of countless technical objects; it is so obvious that it goes unnoticed in the rich Western world, where we only have eyes for much more sophisticated technological promises. It is only when the electricity supply fails that its users can measure the extent to which their world depends on this energy. A breakdown makes tangible the connections which somehow become present through their absence and the lack that it causes. When they fail, interdependencies capture the attention. It is also interesting to note that in recent years we have seen similar experiences with certain digital services offered by companies now known as digital 'giants' (first and foremost Google and Amazon): during spectacular breakdowns, we realized how essential they had become to our practices, professional as well as personal. When we discover, during an outage, the number of services that rely on the infrastructures that these operators have built (such as data centres), or on the infrastructures on which they themselves rely (such as telecom-munications networks), we sometimes have a kind of epiphany. It's hard not to recognize that these breakdowns have educational virtues. How can we better help children understand what the 'Internet' is or, if we are more ambitious, the difference between the network provided by the family Wi-Fi terminal and the 4G telephone service, than thanks to a breakdown? The ease with which a video game can be purchased and 'installed' on the living room console, and the fluidity of remote exchanges with friends, draw on a whole background of telecommunications infrastructures which is revealed when it breaks down – after the initial irritation. Regarding the way a breakdown brings into focus the technical networks in operation, Geoff Bowker and Susan Leigh Star speak of 'infrastructural inversion', an approach which tears users away from the transparent obviousness of ordinary functioning and brings to the fore material aspects usually left out of political discussions.[13] These interdependencies are obviously not just technical. Certain breakdowns also provide the opportunity to discover or rediscover the fact that a particular technology is based on fragile environmental conditions that can be undermined by climatic upheavals (the flow of a river, moderate external temperatures, soil which is not too arid, etc.). In these cases, the political force of the revelatory effect is great since, as well as highlighting poorly known industrial-economic mechanisms, it makes it possible for us to grasp clearly the close entanglements between what we believed could be separated into two worlds – that of technology on the one hand and that of nature on the other.

The revelatory virtues of breakdowns and accidents are sometimes repeated identically in works that extol the virtues of repair. This is the case of Matthew Crawford's *Shop Class as Soulcraft*,[14] which presents the act of repair as a privileged moment of contact with automobile mechanics, one that reconfigures the relationship between a machine and its user. By becoming a repairer, the user learns to understand the functioning of the machine and thus becomes emancipated from the various social, technical and economic mechanisms which immobilized them in the position of a docile and passive user. As a direct extension of the dysfunction that it seeks to correct, repair can thus be considered an epistemic operation in its own right, and seen as a political gesture if it is accomplished by people who do not have any official responsibility. But while it is very effective in freeing people from their fascination with innovation, the conceptual and political seduction that the notion of repair exerts nevertheless runs the risk of reducing one's capacity to take into consideration the range of relationships that can be established between humans and their material environment.

If we want to measure the contribution an approach based on maintenance can make in relation to the revelatory dynamics of breakage and restoration, we need to look at two aspects of this relationship, already highlighted by Ukeles' first performance: routine and the involvement of people 'without quality'. Unlike repair, maintenance requires on the one hand that we be concerned with what does *not* constitute an event, and on the other hand that we take into consideration all those who participate in it – people who are neither heroines nor heroes.

A daily pulsation

In 2015, John Oliver devoted part of his show *Last Week Tonight* to the question of infrastructural maintenance in the United States, which – as he showed with the mixture of in-depth investigation and caustic humour that characterizes his work – suffers from a worrying lack of investment. He concluded his discussion, which was peppered with numerous examples, by expressing regret that maintenance receives much less attention in the media and, more generally, in popular culture, than major breakdowns and other accidents, the repair of which is the subject of ever more spectacular fictional depictions. He proposed to remedy this situation by broadcasting the trailer for a fictional feature film which would focus entirely on maintenance and would be staged on the model of the best 'disaster' movies. The sequence, in which several A-list guest stars make an appearance, is hilarious. We see a supervision team passionately debating how to deal with a small leak, and an operation to tighten a bolt is staged like a bomb defusal, suspenseful music and profuse sweating included.[15]

As always with John Oliver, the comedy is the result of a keen eye for the situation, and arises from an in-depth technical and political analysis. Here, the demonstration is extremely effective: unlike acts of repair, which can be staged as acts of salvation or indeed redemption, maintenance interventions are far too ordinary and boring to arouse the public's curiosity and, even more, attention. Why? Because breakdown and repair leave a clearly identifiable mark over time. They interrupt the course of the narration and thus give the plot a sense of direction. As Oliver shows in another sequence, the breakdown and repair of infrastructure have the same capacity to excite us as design and manufacturing. Maintenance, on the contrary, is far from exciting, he points out, imagining what a parent would say if they gave their child a box of Lego that was a 'maintenance kit' for a bridge: 'It comes built. And then you maintain it, and if you do it right, nothing happens. And eventually, you die.'

Nothing happens. This is a key element of what separates maintenance from repair. As we have seen, if accidents or breakage are major tools of critical analysis, it is because they interrupt the ordinary course of things and initiate a change of state. They are an event. Putting an end to this interruption, the exceptional nature of which it confirms, repair is the final punctuation of this event. It is the operation by which things 'return to normal' and the relationships between technical objects and their users reassume an acceptable, 'normal' trajectory. This process is obviously never presented as perfectly linear. On the contrary, repair is generally described as an action which is not neutral and by which things are never put exactly back in their original place. But they are two states which are highlighted, on a model closely inspired by the Heideggerian dichotomy. On the one hand, we have a 'healthy' state characterized by a peaceful relationship between humans and things – a balance which does not really deserve any attention since it does not pose any obvious problem, or in any case does not offer satisfactory support for analysis or criticism. On the other hand, we find a broken-down, wobbly state, a sort of gap in the progress of the world that requires us to look into it attentively, from the diagnosis of the initial break to the restoration of a new form of equilibrium. This dynamic, which stages an oscillation between two distinct states, can be seen in the very prefix of the word 'repair', which marks an iterative movement. 'Repair' is taken from the Latin verb *reparare*, which means to prepare again, to rehabilitate or restore something. We also find this same idea in the related notion of resilience: in the seventeenth century, the adjective 'resilient' referred to the ability to bounce back, to have a high resistance to shocks. Resilience is this quality of reaction to disruption.[16]

Thus, the vocabulary of repair, like that of resilience, simultaneously emphasizes the break – the event of the shock – and the one-off operation which corrects its consequences. From this point of view, it deviates only marginally from contemporary narratives of innovation, which are entirely focused on the promises of disruption. If it is obviously not a question of using the idea of repair to glorify the process of destabilization itself, it is nevertheless the same interest in discontinuity which is at work and, consequently, the same lack of interest in continuity. By focusing on the virtues of repair, it is easy to leave aside the apparently routine and low-stakes situation before the breakdown occurs and after the reparatory intervention. This is precisely the fabric of this ordinary situation that comes to the fore if we take maintenance activities into consideration: what is not an event but, on the contrary, business as usual, those small, unimportant gestures which neither break nor restore the order of things, and yet play a full part in their existence and are even vital to the stability of the relationships that humans maintain with most of them – as on all those occasions when someone wipes clean a glass display somewhere in the world.

In a text soberly entitled 'Rethinking Repair', Steve Jackson underlines the relevance of detaching ourselves from the restrictive vision of repair which fuels a binary reading of the relationships between humans and technical objects.[17] Actually, more than rethinking repair, he suggests that we mainly need to rethink the nature of breakdown, which should no longer be considered as a provisional and exceptional state, but rather as a permanent condition, always present in the material fabric of human societies. Designers simply need to leave their creative studios, and researchers simply need to lift their heads from their books and shift their gaze a few metres during their investigations, to realize immediately that breakage is everywhere – it's the norm. And it is a matter of some urgency, writes Jackson, to rid ourselves of any thought pattern that presupposes the existence of an order, a balanced state of operation, or even a perfectly aligned use, which are all prerequisites for breakage – states to which each repair operation allows us to *return* with just a few small adjustments. Going against this still omnipresent reflex, Jackson calls for the development of 'broken world thinking' in which neither breakdown nor repair is an event since they do not produce any real break in a relationship between technical objects and their users – a relationship that is no longer considered to be peaceful by default. Thus, Jackson explicitly distances himself from Heidegger's model. Rather than starting from a clear difference between two forms of being-in-the-world, one opaque and problematic, the other transparent and anodyne, 'broken world thinking' generalizes the former and makes itself sensitive to its many variations in an approach which still acknowledges the epistemic virtues of breakdown and repair, but is more inspired by pragmatist philosophy and the central place it gives to inquiry. Admitting that the world is 'always broken' means agreeing to take seriously the gropings, the tinkering, the aborted attempts and, more generally, the disturbances that constantly crop up on the surface of the relationships between human beings and technical objects. The challenge here also involves recognizing an ethical dimension in the constant explorations to which these disturbances give rise. In its extreme version, 'broken world thinking' sees each gesture which contributes to maintaining a tool in working order, and each intervention which helps to make an object last despite wear and tear, as a way of cultivating a permanent inquiry that takes nothing for granted.

Surprisingly, while he continues to depict activities of daily maintenance that have little to do with the idea of a 'return' (which, as we have seen, is central to the etymology of the term), Jackson remains attached to the vocabulary of *repair*. This is why we mustn't relapse into nominalism: it is vain to focus solely on the words used without questioning what those who deploy them are actually trying to highlight. As we will show through the cases that we will be presenting as we continue our explorations, maintenance and repair studies that have developed in recent years are teeming with texts that do not draw any conceptual difference between repair and maintenance. Much more than the word itself, therefore, what matters in Jackson's stance is his invitation to broaden the focus and to stop using the prism of disruption which limits repair and breakdown to the exceptional and clearly demarcated timeframe of the crisis. The world is always broken, and we spend our time fixing parts of it: this is Jackson's message.

This powerful idea that maintenance is *constantly* taking place was developed in wonderful ways in Ukeles' performances. They always contrasted the exceptional character of the creative gesture (and also, one might add, the idea of a masterful restoration following a spectacular breakage) with the daily rhythm of maintenance, an activity that is always being restarted, and that strictly speaking has neither beginning nor end. The many performances on this theme include *Washing/Tracks/Maintenance: Outside*, produced in Hartford a few days after the staging of the 'transfer' that we described above. This time, the performer took over all the spaces of the museum, even the exterior staircase of the main entrance to the building. Kneeling on the floor, she spent eight uninterrupted hours (the length of a standard working day) washing the floors of the exhibition rooms, the main hall and the marble steps, equipped with a bucket, a mop and some of the very special rags that curators usually employ when they delicately dust works of art. Among the things that this performance brought out was the highly repetitive nature of the intervention. Every time someone entered or left the museum, the floor, still wet, was stained with traces. To erase them, Ukeles went over them tirelessly with her mop and rags. The lesson is clear: maintenance work is endless. It never becomes an event, it is not organized around a disjunction between two states of the world, it unfolds in the interstices of days and nights, a time when nothing seems to happen. It is not the exception to routine, the deviation from the linear trajectory of the world: it is that which is continually carried out so that routine and linearity are possible. It generates continuity; it cultivates it. While repair is carried out in exceptional spurts, maintenance is a daily pulsation.

In this sense, it also offers a striking contrast with contemporary discourses which focus on the acceleration of social times, whether these discourses are enthusiastic or anxious about the transformations of the world. Steve Jackson has also insisted on this aspect, showing how these grand narratives neglect the innumerable activities which ensure the proper functioning of technologies on a day-to-day basis – technologies that are supposed to be both the source and the symptom of the accelerations in question.[18] The temporalities of these activities differ profoundly from those which are presented as inherent to technological innovation. Bringing maintenance out of the blind spot that these discourses generate and revealing the temporal flows that drive this maintenance help to nuance, or even frankly destabilize, many of the postulates of what Jackson calls the 'modernist stories of speed and technology'.[19]

As opposed to the linear time of a technological evolution depicted with the broad brushstrokes of unambiguous qualities and associated with the idea of a continuous movement that only major catastrophes can change, the rhythm of maintenance comprises a series of punctuations, of sometimes fleeting moments that accompany the life of objects and repeat themselves irremediably in a monotonous tempo, being never completely different and never completely identical. Over the course of their repetition, these moments form a beat whose rhythm oscillates between the regularity of planned operations and the perpetual surprise of an unpredictable damage that must be taken care of immediately. Jackson uses an organic image to mark both the gap between the two temporalities and the invisibility in which these activities are concealed. He depicts them as the 'slow underbelly' of modernist tales. What happens in this slow underbelly, he explains, completely goes against the teleological vision of increasing acceleration posed as a generic and universal principle.

If we were to represent this fragmented and irregular time of maintenance graphically, made of well-established time loops and incessant improvisations, it would form several lines of flight tangential to the technological trajectories described by those who seek to depict the progress of innovation by flattening it out into a straight line. It would also produce dissonances in the temporalities that most users experience on a daily basis. In a study that lies midway between art and anthropology, Hilary Sample emphasizes these deviations by looking at the history of architecture and its relationship to maintenance activities.[20] She shows how different is the image of the frozen, massive building – both an architectural manoeuvre in the history of heritage and a stable material environment for those who inhabit it – from the incessant agitation of the guards and the cleaning and gardening staff. Changing a faulty light bulb, sweeping the common areas, repainting the stairwells every five or six years, replacing tiles, filling the gaps in sheets of corrugated iron, clearing gutters, unblocking pipes, oiling stiff radiator taps, replacing leaking taps, changing washers on water mains inlets… The list of operations is immense and weaves the fabric of an alternative time which seems foreign to the times of the architects and the residents, even though it constitutes their very condition of possibility.[21]

Another and even more striking example may help to understand the extent of the difference. As one of the exceptional consequences of the pandemic triggered by the circulation of the coronavirus from the end of 2019 onwards, many countries adopted more or less long periods of lockdown which gave rise to sudden transformations in the use of information technologies, allowing people to work or study remotely. In a few days, those who exercised a compatible profession (i.e. mainly an office activity) switched to a regime of digital immediacy. Some of our family spaces (kitchen tables, desks, beds and sofas) were transformed in just a few seconds into professional communication zones via our computers, tablets and phones. Among the countless comments that sought to make sense of this shift – a shift that would have been unimaginable a few months earlier – most concurred in their diagnosis of a dazzling acceleration, one that was both an outcome and a cause of the almost complete interruption of mobility. This acceleration first affected daily activities, disrupting their rhythm. Many of us who did not have to go out to earn a living experienced the sudden disappearance of transitions: the transition of going down corridors and taking breaks around the coffee machines, which provided a very useful airlock between meeting rooms and our offices; that of the street and neighbourhoods through which we had formerly travelled every morning and every evening, as a backdrop to the daily metamorphoses of individuals into professionals, of children into schoolchildren and back again. This impromptu advent of immediacy was understood as the result of another acceleration: that of an almost natural technological evolution which saw companies, administrations, schools and families 'adopt' powerful material and digital 'solutions' capable of delivering on the promises of the digital revolution. So, paradoxically, during the parenthesis of this lockdown, everything went faster, both the march of innovation and the pace of our personal and professional lives.

The specious nature of these descriptions is clear – particularly in their claim to generality, which neglects the profound inequalities of access to the technologies in question and the differences in the extent to which they can successfully be operated. But we tend to forget the extent to which these descriptions give a truncated image of what they classify as technologies when depicting how most people were flung, almost without friction, into what had seemed, just a few months earlier, to be the 'future'. In particular, the endlessly repeated list of the new software that was successfully created in response to the crisis fails to address the layers of infrastructure which allowed the information generated by this shift into the future to be translated, transported and stored. These infrastructures of immediacy remained a blind spot in both factual descriptions and political commentary. Even during the numerous debates which focused on the importance of skilled trades as indispensable, we heard nothing of the people who strove to make them work. And yet, it is easy to imagine how the brutal reconfiguration of uses required additional supervision and extra interventions at all hours of the day and night, not to mention the management of spare parts disrupted by a global logistics that had been brought to a standstill.[22] The rhythm of these activities has little to do with the usual ways in which the pandemic was tackled. But the latter do not exist without the former. The compressed time of our lockdown lives relied on the fragmented and multiplied time of the occupations of maintenance thanks to which the cables, connection points, distributors, relay antennas and even servers stacked in data warehouses managed to do their jobs.

It is to the rhythm of infinite repetition that maintenance alerts us – to this 'eternal return', according to one of the expressions of Ukeles' manifesto, much more than the one-off idea of repair, with its prefix emphasizing the reaction to the interruption. There is also a fairly simple way to illustrate what differentiates the two areas in this regard. Just look at how the two verbs can be conjugated. Saying that something 'has been repaired' suggests that it can, for a significant period of time, carry on without a new repair being necessary. It is precisely this shift into the past participle which characterizes the success of a repair. If it is necessary to repair it again before too long, it means that the first intervention failed and that the thing was not really repaired after all. The situation is completely different on the maintenance side. It is of course entirely possible to say that something 'has been maintained'. But the consequences of such a statement are very different from those of the previous one. It is certainly essential to determine whether or not something has been taken care of in many situations (e.g. to assess the means deployed to prevent a bridge from collapsing, or a boiler from exploding). However, a positive response does not under any circumstances authorize the interruption of the activities in question. While the success of repair is measured by the absence of its reiteration, that of maintenance is, on the contrary, due to it being endlessly recommenced. Something that has been maintained must continue to be maintained, again and again. Maintenance is conjugated in the present. However, this does not mean that it is the same as immediacy – quite the contrary, since it is entirely focused on the need to make things last. As Pierre Caye magnificently puts it, 'Constructing duration consists [...] in transforming the moment into now, in passing from the moment, that is to say the *in-stans*, etymologically that which does not hold, that which is not stable, into the now, into the *manu tenere*, into what we hold firmly in hand so as not to let it escape.'[23]

Obviously, this is not to say that we are not interested in repair. The distinction that we are proposing, by forcing things slightly, simply aims to foreground maintenance, both as a type of activity which goes well beyond the scope of repair, and as a specific concern: 'to make things last', rather than 'to put things in order'. Concretely, this simply means that we will avoid isolating moments of repair but instead approach them in the light of the incessant iterations of maintenance, as forms of punctuation among others.

Neither heroes nor heroines

The exercise that John Oliver and his team engaged in, imagining what a trailer for a non-disaster movie might look like, not only illustrates the gaps between the exciting times of breakdown and reconstruction on the one hand and the apparently monotonous rhythm of maintenance on the other. He also puts his finger on the fundamental immodesty of the restorative actions which, in real disaster films, are carried out by characters with almost mythical features. It must be admitted that putting things back in their place is not an easy task. For the state considered normal to be restored, for the breakdown not to be permanent and for order to be brought back, it is often necessary to struggle. Sometimes you even have to be prepared to fight. In her book, Spelman describes this struggle as the destructive part of repair. Repairing is an arrogant gesture, she writes, which aims to destroy 'the state of brokenness'.[24] Putting things in order means killing disorder, or at least considerably reducing its scope. It is in this war against disorder that the tendency of repair to be spectacular in nature is rooted: it is tempting to present this as the delicate shift from one state of the world to another. And, based on the model of disaster films, it is easy to see the figure of the repairer (very often a man) as a hero who, by dint of tacit knowledge and ingenuity, and often physical strength, manages to make some indispensable object or some essential infrastructure work again, and thereby ends up restoring the sociomaterial order which rested on these things. There is of course nothing fallacious in this heroism. We can probably all remember some life-saving intervention at the end of which we seriously felt that the person who had managed to restart our computer, repair our boiler or restore the electricity supply to our home after a storm had, if not saved our lives, at least done us a great service.

We have no intention of denying the existence of these heroic figures. On the other hand, it is important to take stock of what they can hide. Just as the great stories of innovation only retain from history the names of great men who are considered to have 'made a difference', an exclusive focus on the breakdown/repair dynamic (or its counterpart, crisis/resilience) runs the risk of isolating a few exceptional figures, people who succeed in putting the world back on its feet, while leaving in the shadows those who, day after day, strive to prevent breakdowns from occurring by carrying out countless small operations, for the most part anodyne, which make it possible *not* to make a difference.

What if we gave these people the central role in our stories? asks John Oliver. What if we tried to populate our stories in new ways by correcting the focal length, but sometimes also by moving the camera sharply to avoid certain off-camera effects? Who would we then bring into the frame? Many, many people, replies Ukeles. The maintenance population is immense and heterogeneous. And if we wish to take the measure of this multitude, we must abandon the idea of spotlighting the isolated figures of a few men (and even fewer women), who would simply be added to the list of heroes and heroines. We must in fact turn the question around and seek to identify the type of maintenance that each person carries out in the most daily aspects of their lives. With this in mind since the very first collective travelling exhibition to which Lucy Lippard invited her in the early 1970s, Ukeles has been drawing on many questionnaires, in-depth interviews, observations and photos to lay bare this incessant work accomplished by her fellow artists, by the maintenance staff of artistic institutions and by the crowds of anonymous people who come to visit her exhibitions and attend her performances.

In 1976, on the occasion of her performance *I Make Maintenance Art One Hour Every Day*, she invited 300 support services employees from an office building to the Whitney Museum, after having spent their entire shifts together with them, day and night, over a period of five weeks. Isolating an hour from each day, she asked them to describe their actions, letting them choose between 'maintenance work' and 'maintenance art'. She recorded each response in the frame of a Polaroid taken during her observations, and the photos were hung in the museum one by one, until they formed three huge panels made up of 690 photos divided into two-thirds nocturnal activities and one-third daytime activities. The revelation proved to be double-edged: the workers suddenly found themselves at the centre of attention of those who visited the exhibition and discovered what lay behind daily life in the office building in question. But by fully participating in the performance, these women and men of maintenance also saw themselves exhibited in the museum space, and realized that their actions could be staged and recognized as those of artists in their own right.

This repopulation has a striking effect. It is also found in *Touch Sanitation*, arguably Ukeles' most famous performance, which marked the first years of her permanent (and unpaid) residency in the New York City Department of Sanitation. Between 1979 and 1980, for eleven months – much longer than she had envisaged when initiating the operation – the artist followed the employees of the department in their urban wanderings, from the city streets to the landfills, from the offices to the locker rooms. Throughout this performance, she shook hands with all 8,500 New York City garbage collectors, rewarding them with the same phrase at each handshake: 'Thank you for keeping New York City alive.'

By discovering the techniques developed by Ukeles and particularly reading the prep notes for this performance, one can only be struck by the similarity of her approach with anthropological or sociological research.[25] While she begins by composing an immense map of the maintenance of the city, and then concretely plans the operation, she produces a fully fledged urban study, with both political and epistemic repercussions. It isn't just art that is at stake here, but the full recognition of a swarm of maintenance workers who are barely ever considered by city officials and inhabitants. In the letter presenting her project, she directly addresses these workers, explicitly highlighting their skills and their knowledge of the city: 'You are probably the biggest experts on what's going on around here. I'm amazed when I talk to sanmen. They can tell me what a street will look like even before we come there. Magicians. Of all the people who symbolize the City as "public workers", you know the City. You feel it in your hands, in your backs, in your bones.'[26]

Over the course of her long-term performance, Ukeles seeks to show this expertise, to get people to become aware of it. While she accompanies the garbage collectors on their rounds, she also imitates their gestures and strives to highlight actions and know-how that go completely unnoticed in ordinary urban flows and are not highly valued in comparison to what other New York City employees do – things that everyone agrees are of crucial importance. Her letter continues by explicitly putting her finger on this question, in a paragraph which alone sums up what we are trying to formulate in this chapter: 'More than the policeman and more than the fireman. *They're* there to handle things gone wrong and out of commission. *You're* there to handle the "normal", what's going-going-gone, what keeps coming continually.'[27] These few heroes and heroines deal with extraordinary situations. The numerous maintenance workers take care of normalcy. And it's one hell of a job.

However, it would be simplistic to limit Ukeles' work to the repatriation, into the field of art, of categories of people from professions and workstations explicitly associated with maintenance. As we mentioned above, her ambition is much greater. Starting from her own activity as a mother taking care of her children and her house, and combining the domestic sphere, offices, gardens and the city, her work weaves practical solidarities, both bodily and intellectual, of a completely new kind. Wiping the table, washing the dishes after a meal, cutting your nails, airing the room, taking out the trash, picking up the trash, checking the oil in the car, maintaining the asphalt on the road, reinforcing the piers of a bridge, putting on a new layer of coating, renovating a facade, inspecting the rails on a railway line, the bogies of the train carriages, or an aeroplane cabin, checking the integrity of a database... Everyone carries out maintenance tasks, even if many never realize it. Rather than heroes and heroines who can be identified once and for all, what is important to highlight is the variety of forms of care that are given to things, just as much as the concrete conditions which allow the women and men who provide them day after day to remain neglected.

Reassigning attention

Let's recap. Placing maintenance at the forefront opens up an analytical horizon very different from favouring the decisive breakdown understood as a significant event which interrupts the ordinary course of the world. Here, routine takes on a much less bland aspect, since its repetitions constitute the essential engine for the generation of continuity. And it is those who care about things that we need to consider, rather than the few people who have been turned into heroes by putting the world back in order. That said, the emphasis placed on maintenance does not eclipse either innovation or repair. We have known for a long time that innovation and maintenance are closely linked. David Edgerton, for instance, highlighted how many innovations have been developed within repair shops themselves.[28] By apprehending them from the point of view of the uninterrupted rhythm of maintenance, we are simply suggesting a new way of describing innovation and repair. Without losing relevance or singularity, they are stripped of the exceptional character on the basis of which they are usually considered.

Following the paths opened by Ukeles and Oliver, it is therefore a *reassignment of attention* that we propose to carry out in this book. Let's take a simple example to summarize what this shift consists of, one that will help to rid ourselves of the reflexes of the master narratives of innovation, and to extend certain framing effects created by repeated calls for repair. Let's imagine that we wanted to account for the contemporary use of the bicycle. It is no secret that cycling has assumed a new importance in many cities, including those which seemed to have turned their backs on it since the advent of the car. There are a thousand and one ways to proceed with this description, depending on who wishes to do so and the issues they seek to address. If we want to place innovation at the forefront and highlight the current buzz found in this sector of activity, nothing could be simpler. We need merely emphasize the dazzling advent of electric bikes and smart bikes, which, one can argue, have profoundly transformed many activities by promoting daily home–work travel, or by associating new services with mobility such as navigational assistance or the generation of useful data both for cyclists and for municipalities wishing to better understand the travel habits of their inhabitants. If we are concerned about city policy and urban planning, we could also turn to infrastructure and highlight the recent initiatives of municipalities which have increased the number of facilities dedicated to cycling, radically changing the cityscape of many urban areas and promoting the acceleration of decarbonization in the realm of transport. The list could obviously be extended.

Without denying the relevance of these approaches, we may also want to highlight other aspects that contrast with their technophile inclinations. We will then factor in the importance of the question of repair activities in cycling. Cycling isn't just about riding a new bike down brand new trails. Think about the effort it takes to disassemble a tyre, take out the inner tube, discover the location of the hole, correctly install a patch, then reassemble everything so that each element returns to the position that allows you to ride safely. Not to mention the skills needed to repair a crankset or replace the gear shift housing. Another side of the story then comes into view – one that shows the many possible events that act as obstacles to cycling. Continuing this line of argument, we can show that the know-how necessary to remove these obstacles is not equally shared among the population and that we should also be concerned about the ecosystem of the workshops and the technicians who play a crucial role in the dramatic spread of the practice. This, indeed, is what French government officials did in the early stages of the Covid-19 pandemic, offering a fifty euro 'repair' cheque so that those who wanted to travel by bike rather than by car or public transport could repair their bicycles at a lower cost. The concern to paint a less arrogant picture also applies to infrastructure. The unprecedented crisis caused by the Covid-19 pandemic, for example, gave certain cities the opportunity to demonstrate remarkable resilience by adapting part of their public space in just a few days. In Berlin, Bogotá and Paris, but also in many less emblematic cities, kilometres of temporary tracks were installed in record time, providing the necessary conditions for city dwellers to resume their journeys on bikes.

Finally, without exhausting all the possibilities, we can draw inspiration from Ukeles' performances and Oliver's fierce humour to further broaden the scope of the description. In this case, we will start out again from the paths based on issues specific to repair and resilience, adding to them the host of apparently insignificant interventions whose recurrence nevertheless ensures a certain sustainability for both bicycles and the infrastructure dedicated to them. Oiling the chain from time to time, pumping up the tyres regularly, tightening the brakes and monitoring the condition of the brake pads, checking and readjusting the battery fastening when it gets damaged by vibrations, recharging the battery at the right pace to optimize its life... Monitoring the state of the roads, filling the cracks before they form potholes, pruning the hedges which border certain cycle tracks... These are all banal gestures which do not punctuate our stories with any key moment, drama or suspense, but which reveal a part of the world that the first two versions of the story leave in the shadows. They are all gestures that it would probably be superfluous to model or list in some attempt at exhaustivity, but which can teach us a great deal about the material fabric of the world and the richness of the relationships that are woven in it.

2

Fragilities

Our own 'initiation' into maintenance took place during our investigation into the signage of the Paris metro.[1] Begun in 2007, it was part of a research programme which brought together colleagues from different disciplinary backgrounds and aimed to investigate the political dimensions of writing, in a wide variety of fields.[2] In particular, we wanted to question the power of writing, its capacity to concretely organize our ways of living together. In this respect, metro signage presented several kinds of interest. A seemingly innocuous graphic device, it is in reality a powerful operator for ordering space and behaviour, a crucial element in the fluidity and safety of urban mobility flows. During the first months of our investigation, we gradually identified what we considered to be the main sources of power of this network composed of countless scriptural artefacts. If the Paris metro signage was so effective, and if it was considered to be one of the most successful in the world, this was because it had been at the centre of an ambitious effort at standardization. In the 1990s, in fact, all of the RATP signage was redesigned and each of its dimensions standardized down to the smallest detail. From the size and shape of the arrows to the location of the words and each sign on the panels, everything was specified, down to the level of a single millimetre, in a voluminous book of graphic instructions. Each name and each abbreviation were established, as was the set of colours calibrated to identify the metro and bus lines, as well as the tramway. The signboards themselves were organized into modules of specific shapes and sizes, aligned with the iconic tiles of the Paris metro. Precise rules were established to identify their number and their position in each space of the metro, from the entrance lobby to the platform, including the corridors and staircases. There was no doubt in our eyes: in this policy of graphic homogenization, we held one of the main keys to the power of signage to create order.

In parallel with these first results, we saw the actual material of the signs as the other crucial element of this power. Among the people we met, many insisted on the importance of enamelled sheet metal, whose unfailing solidity ensures the sustainability of the way as a whole. The material is the secret which allows the signboards to resist repeated cleaning. It provides a valuable complement to the extreme standardization intended to ensure a graphic consistency: the longevity of each sign, without which the power of the scriptural assemblage would be largely minimized.

At this point in our investigation, everything was relatively simple. Those particular graphic objects, the signage boards, had been invested with a role which had less to do with their 'informational' mission than with certain graphic and material properties. Designed to be the same everywhere and to last, they functioned as a robust system, a fixed infrastructure dedicated to mobility, in the same way as the architecture of the metro stations. In simple terms, we'd turned into fetishists, fascinated by this ability of signs and their carefully designed graphics to compose a world where it would be vain to try to separate the linguistic from the material dimensions.

As our understanding of this stable graphic device itself grew in solidity and rigidity, our investigations seemed to be coming to an end. One innocuous remark, however, was enough to arouse our ethnographic curiosity, and then, dangerously but fortunately, to cause our analytical edifice to topple over. In fact, it was the person responsible for the standardization of all RATP signage, the same person who had given us access to the different standardization components of the system, and told us part of the story of its design, who caused this unexpected rethink. At the end of what we thought would be one of our last meetings, as we went through the list of additional people we could possibly interview, just to complete the picture before concluding our investigation, this person mentioned that we could perhaps go and see the 'maintenance guys'. The suggestion hit a nerve, and we jumped at the chance. The idea was all the more appealing because it could give us the opportunity to practise the research method that suited us best and that we had not been able to practise until then in this project: direct observation. In addition to in-depth interviews and the examination of archives and technical documents of all kinds, we would perhaps be able to add a few hours spent with the workers directly in contact with the signboards. The trail proved to be fruitful – well beyond our expectations – since it led to several days of observation, in the maintenance workshop and in the metro stations. Above all, it contributed to profoundly transforming our understanding of signage.

The head of the signage maintenance department welcomed us with pleasure, shortly after our conversation with the standardization manager. He agreed that we should join the intervention teams for a few days. We were convinced that we could not really understand maintenance without going through a form of immersion in the daily life of the department, but we were far from imagining the importance of what this immersion in the most ordinary and most repetitive type of work would help us to discover. We will have occasion to return several times to the activity of these maintainers: their gestures, their vocabulary, the practical details of their interventions. At this stage, we will stick to the fact that the discovery of their work changed our view of the signboards, these graphic objects that we had until then looked at – outside our own travel in RATP spaces – from the point of view of the people responsible for their design.

Our first visit to the workshop was enough to alert us. While, on the computer screens, we found the elements of the highly standardized nomenclature of signage, and while we spotted a few copies of the graphic manual available to the agents, we also realized that everything wasn't working as simply as we had imagined. On this occasion, we discovered that some of the signs at the station were 'temporary' PVC signage boards. Printed, cut and laminated on site, they were intended to replace defective signboards while waiting for their final version, in enamelled sheet metal, to be delivered. But it was by going on a 'round' and following in the footsteps of the operators working at the stations that we saw a radically new world open up before us. The signs – which we thought we knew in every detail – were literally transformed before our eyes. After a day when each of us followed a pair of maintenance workers, we found we had already encountered a considerable number of situations in which these emblematic graphic objects appeared to us in states that we had not suspected. One of them, in its provisional PVC version, had buckled across a good third of its surface and had cracks that distorted the letters and lines there. The colours of another had faded considerably, despite the underground environment, and were now reduced to just a few shades of blue. Yet another, a rather large one in enamel, hung dangerously from the wall and seemed as if it might come loose at the slightest draught of air. Others had significant scratches which constituted what maintainers call 'scratchities', a particular form of graffiti engraved on the surface of the objects using a key, a screwdriver or a penknife. Among the problems that these first rounds introduced us to, humidity was undoubtedly one of the most striking. Many stations are in fact subject to sometimes significant seepage. When the signboards are exposed to water, they deteriorate quickly and show traces of mould or even rust. As we followed the agents, we discovered a considerable number of signs affected by these liquid attacks. We also learned that action was sometimes needed because various signboards had simply disappeared, probably taken away to be sold to a few enlightened collectors.

If for now we concentrate solely on the material components of signage, the objects which compose it and the way in which they appeared to us during these observations, it is an understatement to say that the first interventions at which we participated changed our vision of things. From the users' point of view, signs are part of the graphic and architectural landscape that constitutes the world of the metro. They are ubiquitous points of support, and a simple glance is enough to use them when moving on familiar terrain, while more sustained attention helps to identify the direction to follow if we discover a line or an interchange hub. Rare are the cases where the signage system is inadequate. Essentially, this happens solely during major work during which the stations are completely dismantled, so that a real lack is felt and we realize the extent to which these innocuous objects help to guide our movements. But we ourselves could no longer be considered ordinary users. After several months of investigation, we had somehow reinforced the way travellers on the metro view these signs, by discovering the framework of rules and standards on which all the graphic components were based and organized into a system to form the Paris metro signage. But neither the position of a user, nor that of a researcher sensitive to the dynamics of design, enables one to understand these objects in the light of maintenance activities. Until then the signboards had seemed robust, immobile, resistant and always available; now they revealed themselves, through the filter of maintenance operations, as vulnerable, mobile and subject to perpetual transformations. They had seemed solid on the pages of the graphic charter and the studies which had preceded their installation, but in situ they appeared to us fragile, likely to deviate drastically from the stable version that is guaranteed by the standards which precisely define each of their characteristics once and for all.

The sessions during which we shadowed the signage maintainers led to repercussions that were important not only to our investigation, but more generally to our trajectories as researchers. They led us to explore maintenance from very specific questions. Rather than approaching it as just one occupational site among others, in which it is essentially work relationships that are at stake (as research in sociology of labour and organizations studies had already done),[3] these observations led us to approach maintenance operations in their specificity and to question the productive role they play in activating the force of the wayfinding system. We very soon realized that what we had seen in the stations while following the maintainers did not actually contradict the first avenues that we had explored. There was not, on the one hand, a 'myth' of graphic standardization and material solidity and, on the other, the 'reality' of damaged and changing signboards that had to be taken care of on a daily basis. It was all part of the same reality that we had discovered. A dimension to which most users generally do not have access, but which plays a full part in the graphic ordering apparatus. In a way, we had simply switched from a *semiotic* model, thanks to which we were able to isolate a few components that defined the force of signs, to a *pragmatist* model, which insists on the continuous work necessary for the permanent restoration of this force.[4]

But let's get back to the signs and what their maintainers taught us to see in them. At the beginning of this narrative, we used the perhaps rather strong term 'initiation'. It has the merit of underlining the effect of the days of observation that not only allowed us to understand the activities of the operators and to discover the details of their actions, but also profoundly reconfigured the way in which the semiotic and architectural environment of the Paris metro presented itself to us. This reconfiguration was not very different from the one described by Baptiste Morizot, who relates a hike along the paths of Yellowstone Park, in his book on trails and tracking.[5] In a few lines, he indicates how the traces of the presence of a bear, when spotted by a walker, immediately disrupt the experience of the environment: 'A single invisible bear transforms an entire mountain range, imbuing it with another gleam. He gives relief to each bush, which now has a hidden side. He digs a new depth into the thickets, which regain their dimension as *habitats*.'[6]

In anthropological terms, there is nothing in common, of course, between perceiving the presence of a predator – something which, as Morizot explains, places the trackers in the position of 'one living creature among others', reminding them that they can also become meat[7] – and discovering the material transformations that affect signboards that had seemed immutable. But it is a comparable effect of destabilization, in the strong sense, which is at work: the emergence of dimensions of the world that are ignored because they are literally imperceptible, reorienting our postulates and making our certainties waver. This is the main outcome of the attentional transfer operation that we mentioned in the introduction. Maintenance brings to the surface a constantly changing world of things: cracks in bridge piers, pipe leaks, rust, crumbling surfaces, deformations and abrasions, seepage, mechanisms that seize up, nuts that loosen. A world that contrasts with the world of solid artefacts and unalterable infrastructures on which one can rely. Maintenance works like a prism that places the fragility of things in the foreground, and makes it matter.

The operation works with all kinds of objects, from our front door to our car, including the pavement we walk on every day, the computer we use at work or the gigantic machine tool with which we interact if we are a factory worker. Everything, seen through the eyes of those responsible for maintaining it, reveals itself to be fragile in one way or another. And the more something is presumed to be solid, considered as a stable and inert object, the more striking the effect of contrast is. This is what Tim Edensor explains, using the example of the stone of Saint Ann's Church in Manchester.[8] In a fascinating article, he shows that the restoration of the church in question reveals the fragility of stone, a construction material considered to be resistant par excellence, an emblem of durability and immutability. Once we look at the activity of those particular maintainers known as the restorers of heritage buildings and take seriously the practical problems they face, we can measure the extent to which stone itself is vulnerable and how far it is from being as unalterable as one might believe. Edensor identifies four main causes of transformation to which restorers are particularly attentive. The first involves meteorological conditions. Rain, wind, periods of drought, frost and extreme heat are all elements which contribute to damaging the stone of the church, causing weathering and erosion in certain places, creating or worsening cracks in others. Air quality constitutes a second source of fragility. The presence of too much sulphuric acid and nitrates tends to dissolve the calcium carbonate, which, on melting, forms a layer of sulphate on the surface of the stone, making it sensitive to heavy rain. The use of coal was of course a cause of great atmospheric pollution and for years the walls of the church were covered in soot, which, as restorers recently discovered, had led to accelerated wear of certain stones. A third source of fragility is the cohabitation of stone with other materials. Some types of mortar created almost watertight areas where water built up. Elsewhere, the metals present in the fastening systems used have, over time, caused rust spots to appear in the fabric. Finally, Edensor explains that restorers are also aware of the action of a large number of living organisms, animals or plants, which undermine the integrity of the stone in one way or another. Birds, bats and rodents, as well as bacteria, plants and even algae, all participate in the mutations of the stone. The people responsible for restoration are very concerned about this powerful menace, and know how to identify sites that are more vulnerable than others to the development of algae or lichens.

Normally, of course, it is impossible, unless you have been trained in materials science, to suspect the extent and diversity of the forms of vulnerability of a material like stone. Whether they are those which make up Saint Ann's church or others, stones seem to the majority of people to be examples of peaceful immutability, symbols of longevity, even of the unalterable force of matter. By approaching them through maintenance, through the concern for the fragility that characterizes them, we can become sensitive to the multiple modalities of their deterioration. Those people who organize their activity around these vulnerabilities can thus become our initiators and help us take stock of material fragility.

But what exactly should we do with this material fragility? Why insist on this point, which seems quite banal once we have used a few examples to highlight it? If we feel it is important to dwell on it at this stage of our reflection, it is precisely because, in reality, there is nothing obvious about it. Or rather, the consequences of taking it into consideration deserve to be made explicit, as they are far from trivial. Material fragility first asks us to re-examine what we think we know about objects and to rethink properties that we imagine to be intrinsic. How many times have we been told – by people who urged us to see beyond (or outside of) our imaginings, our beliefs and symbols – that material reality, or even simply the 'real', was what we could bump into; what was there, quite simply there; what, above all, resisted? Maintenance turns this argument upside down. It sees and shows in matter everything that doesn't hold, everything that can be damaged by our bodies and our uses, and can get worn out by other materials to which we generally pay less attention. Maintenance worries about the part of reality that doesn't hold together. Focusing on the fragility of things thus amounts to highlighting the significance of the material existence of the artefacts that many of us all too often neglect, particularly in the social sciences, where these artefacts have long been understood solely in terms of solidity and longevity and have only rarely been questioned as things with an unpredictable becoming.

Societies populated with objects

Towards the mid-1980s, objects made a dramatic entry into the landscape of the social sciences. In the decade that followed, vehement debates about them shook entire disciplinary fields from top to bottom. What place should be given to them in analyses focused on human beings and their relationships? Had the material part of the world not been too neglected in the descriptions and theoretical developments that had hitherto prevailed? Yet, is it reasonable to consider that objects matter in what we call 'society' in exactly the same way as women and men? Do not sociology, history, anthropology and psychology have everything to lose by ignoring the border that they had carefully erected between beings endowed with reason and the inanimate material background against which those beings move? Beyond sometimes pointless controversies, these questions led to significant upheavals in certain areas of the social sciences. Without reconstructing their genealogy in detail, it is useful to return to their most prominent features in order to understand what is changed in the relationship between these sciences and technical objects if we follow in the footsteps of maintainers.

It was a series of works from science and technology studies that led to the first analytical moves. In France, at the Centre for the Sociology of

Innovation at the École des Mines, the investigations of Michel Callon and Madeleine Akrich into engineers and their productions, Antoine Hennion's research into artistic practices and Bruno Latour's essays on ordinary objects crystallized a large part of the issues and debates. In 1994, Latour published an article condensing a large number of the arguments he was at the time propounding with his colleagues, under the provocative title, in French, 'Une sociologie sans objet? Remarques sur l'interobjectivité' ('A sociology without an object? Remarks on interobjectivity').[9] The ambition was clear. It was not simply a question of including a few objects in sociological description and analysis, nor even of focusing on this or that type of artefact (which would ultimately have been quite commonplace), but of completely reconsidering the role of objects in the lives of human beings and, consequently, profoundly rethinking the consistency of 'the social'.

The path that this sociology *with objects* follows is the path of agency. It is by seeking to identify 'that which acts' without dividing up in advance the skills and powers involved between the entities in question that such a sociology can gradually unweave a social fabric whose heterogeneity it assumes. Among the numerous terms used by these authors, as well as by those on whom they rely and those who subsequently contributed to this area of the social sciences, the notion of *delegation* is undoubtedly the most telling. If we keep track of the action, Callon and Latour tell us, we will understand that a significant part of it is entrusted to objects. The examples that have become classics in the field include the barrier which prevents animals from moving away without the shepherd having to worry about them, or the hydraulic door-closer responsible for closing the doors automatically.[10] In fact, you only need to look around you, wherever you are, to see how many actions are delegated to artefacts. Our world is populated by objects that act on our behalf.

In the article that we just mentioned, Latour generalizes the argument and sees this capacity to surround ourselves with innumerable active artefacts as one of the particularities of human societies – particularly with respect to our cousins the baboons. Without looking into the delicate question of the specific character (or not) of the phenomenon, it is the main movement of the analysis that must be retained, which consists of affirming that objects literally comprise a society with us. Indeed, what is at stake in this reconfigured sociology is repopulation.

It endeavoured to highlight the missing masses of the social, which are composed of all those objects that, until then, had been reduced to the roles of extras, sometimes simple mirrors of purely social forces, sometimes arbitrary signs of identification and distinction.[11]

A similar movement has also developed on the fringes of psychology, where authors such as Edwin Hutchins, Don Norman, Jean Lave and Lucy Suchman have helped to redefine the contours of cognition and intelligence, which were areas of huge interest on the part of computer engineers wishing to model mental processes in order to implement them in their machines.[12] Each in their own way, these researchers from varied backgrounds fought against the all-powerful mentalism of the time and set out to undermine the foundations of the omnipotent 'brain', an isolated organ placed at the centre of all mechanisms of thought and calculation. Drawing in particular on Lev Vygotsky and Alexis Leontiev, they too made room in the description of 'psychological' processes for the material environment and the objects which accompany human reasoning and thinking. Rather than focusing on invisible mechanisms confined within the skull, they argued for studying cognition as a hetero-geneous phenomenon, always located in a specific environment and – crucially for their argument – distributed among humans and things. Norman and Hutchins use the notion of 'cognitive artefact' to describe those objects to which are entrusted operations that, without them, would be mainly mental, such as memorization or calculation. From the most basic (sticky notes, lists, diaries, etc.) to the most sophisticated (devices to aid the piloting of large ships or planes, in which Hutchins specialized), these intellectual technologies are part of cognitive systems much larger than the small perimeter of our craniums.

Similar concerns can be seen at work in the anthropology of material cultures, where many authors endeavour to highlight the role of objects in human societies.[13] The challenge here is to break with the idea of the 'symbolic' – omnipresent in the discipline – that empties things of all material flesh, so as to centre analysis on the physical properties of artefacts and their role in the concrete organization of human relations.

Repopulating the world with its missing masses, practising human and social sciences *with* objects while taking into account their own actions: the shift is far from self-evident, especially since it often involves a significant transformation of methods within the disciplines concerned.[14]

But the issue is not only descriptive. Part of the argument also consisted in questioning the nature of these objects' agency. Latour in particular has shown an interest in the fact that certain objects have the power to constrain human behaviour. The action delegated to these particular objects makes things happen (*faire faire*). On the road, the speed bump, known facetiously as the 'sleeping policeman', forces me to drive over it with my car below a certain speed if I don't want to be violently shaken or to damage my shock absorbers. The washing machine door requires a two-minute wait to open after the program has stopped in order to avoid any risk of flooding; the metro turnstile is only unlocked if the traveller validates a ticket or presents their magnetic subscription card; the electric tool refuses to start until the protective cover is back in place; the armrests added to public benches prevent any normally constituted human body from lying down… Thus, certain processes of delegation take the form of a material translation of social or even moral concerns; this gives objects a prescriptive role. It's a sort of transubstantiation of what Deleuze and Guattari call *mots d'ordre*, instructions and orders transformed into stone, metal, plastic or electronic components. At the end of the operation, the world's principles of organization, rules of justice and other conventions have not simply changed form, they have gained strength. For when it comes to prohibiting or obliging people, written or oral instructions issued by an ordinary fellow human being, or by a representative mandated by the state, pale in comparison with the purely mechanical capabilities of certain artefacts. Not only do these objects act, but they are powerful – sometimes even more powerful than the humans in charge of carrying out the task delegated to them.

Clearly, this observation goes well beyond the sole question of the material densification of the social. If certain artefacts – from the most innocuous concrete protrusion to the most complex infrastructure – exhibit such characteristics, it must be fully recognized that they contribute to shaping the world in which we live. As Madeleine Akrich suggests, we need to understand – and question – the political content of such artefacts as active elements in the relations between human beings, and in the relations between humans and their environment.[15] We must therefore abandon the temptation of any clear division between things and people (that pillar of Cartesian thought) and hijack, or generalize, Clausewitz's aphorism. As he famously wrote, war is the 'continuation of politics by other means'.[16] Likewise, Latour and his colleagues claim that the design and use of a large part of the objects that surround us are other means of doing politics – means that are admittedly less spectacular and more peaceful than many others, but no less effective.

However, one question remains unanswered. What gives objects their force? What exactly is their ability to 'do politics' based on? The examples given by the scholars whom we have mentioned so far, and more generally by those who emphasize these 'active' artefacts, leave no room for any ambiguity on this subject: if objects matter in our world, if they play a full part in human societies, it is above all because they demonstrate a certain solidity.[17] They resist, they do not disintegrate at the slightest flick, at the slightest movement in space. They act because they are materially capable of lasting.

All things considered, this is undoubtedly where the 'dose of fetishism' that Bruno Latour deems useful for the repopulation of society resides:[18] in this fascination with robustness, with stability. And this is, of course, where the difference with the experience of maintenance is most marked. There is no room for deterioration, wear or decomposition in stories that seek to show the importance of objects in social life. There are of course a few accidents, a few dramatic breakdowns, but, as we saw in the previous chapter, the use of these exceptional events generally just reinforces the invisibility of maintenance and its challenges. The ordinary vulnerability of objects does not seem to fit with the general argument that consists of highlighting their capacity to act in the mode of resistance and sustainability.

It would undoubtedly be too harsh to assert that fragility constitutes a complete blind spot in these different analyses and of the model of delegation in general. It is more accurate to say that it is neglected. This is mainly what Susan Leigh Star criticized Bruno Latour for, based on his analysis of the action of a door-closer.[19] In her view, Latour is content with a far too superficial treatment of the discrimination caused by this mechanism: while it effectively allows the door to close 'by itself', it does so at the cost of excluding people who do not have the physical capacity to make it open, as this is made difficult by the forced-recovery device.[20] In her demonstration, Star also regrets the very small number of mentions that Latour devotes to maintenance work and his way of relegating it to a secondary status, even though it is necessary for the proper functioning of the technical object to which the action of delicately closing the door has been delegated. She then indicates that there is no reason at all to neglect maintenance.[21] We concur, though her trenchant stance is intended less as a criticism meant to refute Latour's approach than as a plea to broaden the horizon of the arguments that he is putting forward in this text. If we want to question *in depth* the specifically political dimension of objects, explains Star, we cannot afford to miss what delegation actually involves. In particular, we cannot act as if the maintenance work which determines the proper functioning and longevity of the artefacts to which the burden of acting on behalf of humans is entrusted did not really matter. Nothing in the theoretical and political project of a sociology *with* objects justifies this commonplace and uninterrupted activity being left in the shadows.

In her criticism, Star mainly focuses on workers, and the twofold invisibility from which they suffer. Most often working behind the scenes, far from the users of the objects they maintain, they are in fact simultaneously confined to the margins of analyses that emphasize the agency of artefacts and their political force. These analyses, Star asserts, should show this 'human' action being deployed in the shadow of the mechanical action of non-humans. While obviously taking this demand into account, we think we can extend her argument and question another aspect neglected by Latour, no longer regarding the people who work, but the objects they take care of. In line with Star's empirical and analytical approach, we feel that there is no reason to highlight 'active' artefacts while forgetting that all objects, in one way or another, get damaged and wear out.

Actually, an observation of maintenance activities can help us push the argument forward a little further. In order to understand the political role of objects, we can – rather than adding to our descriptions a greater or lesser dose of concern for fragility – reverse the descriptive movement and try to start *from* this fragility, to make it the heart of the analysis. This will expand what we learnt from our study of signage. This is not to deny the material resistance and persistence of objects, of course, but to understand them fully as *things* – as the provisional result of actions taken well beyond their material limits. Solidity, from this point of view, appears as a condition (in the sense of a condition of life), rather than as a property that characterizes the essence of certain artefacts and alone explains their capacity to act.

Maintainers therefore help us to investigate the material part of the world by encouraging us to stray from the beaten track. In doing so, they introduce us in their own way to new theoretical horizons. For some years, indeed, there have been – in philosophy, but also in sociology, in history, in anthropology and even on the margins of archaeology – fascinating debates about the various modes of existence of matter, and what they can teach us about the world that we inhabit and that inhabits us. Tim Ingold has insisted on the important limits of the very vague idea of 'materiality', an idea that has long held sway in the social sciences to signal the importance of objects in the human world; and he has argued that we need to take into consideration the multitude of materials that contribute to the existence of things, above and beyond the all too pervasive figure of 'crystallized' objects whose inertia and uniqueness are quite relative.[22] In his view, this is the way to acknowledge the provisional nature of those artefacts which he considers most of his colleagues take far too quickly for granted. Every object is fragile because the materials that compose it continue to change, and pursue their own transformations. 'Whatever the objective forms in which they are currently cast, materials are always and already on their ways to becoming something else', he writes,[23] echoing the words of Karen Barad, who, for her part, emphasizes what she calls the 'intra-activity' of matter, forever undergoing transformation.[24]

More or less directly connected to this concern for materials are the numerous interdisciplinary discussions that have forced a rethink of the question of matter by linking it specifically to feminist theories,[25] the concerns of environmentalist ethics[26] and the new approaches found in contemporary anthropology,[27] and have also argued for proper attention to be given to bodies, the cosmos, the flows that traverse them and constitute them, as well as the soil, the air, fire, oceans, plants and animals...[28] The work that has contributed to this renewal is extremely varied. That said, these studies all share an empirical and analytical interest in what some have described, for want of a better term, as 'posthuman', or 'more than human',[29] a posture that detaches itself from the Cartesian heritage of the modern sciences by refusing to see the omniscient human being as the measure of the world. More or less directly associated with

Alfred North Whitehead's philosophy, this de-centred thinking is, quite explicitly, speculative. It ventures into sometimes radical otherness (stones, rivers, trees, wild animals, the materials themselves...) to try to describe reality from new points of view, by following the trace of materials that had hitherto been neglected.

Probably because both the present authors were trained in sociology – with, moreover, a leaning towards workplaces and activities – we have no intention to pursue these philosophical arguments ourselves. We believe, on the other hand, that maintenance practices provide, in their great variety, a site of extraordinary richness on which these discussions can draw. When they make contact with matter, those who try to maintain things, to prolong their existence in one way or another, never cease to experience it, to test its strengths and weaknesses, to deal with its variety, its heterogeneity and its mutations. It is thus with the maintainers that we will learn to speculate, discovering that the practical problems that some of them face on a daily basis resonate with the most abstract questions raised by 'new materialism'.

The diplomacy of alteration

As we showed above, the practical and analytical approach that consists of taking material fragility as a starting point is particularly striking when it concerns objects that we are used to considering as 'naturally' solid: the stone that Tim Edensor speaks of, the highly standardized signboards in the Paris metro... Rotor, an interdisciplinary collective based in Brussels, has magnificently demonstrated what is at stake in such a reversal by devoting one of its productions to the question of wear and tear. Initially specialized in projects involving the recycling of construction site materials and the circular economy in architecture, the members of Rotor joined forces with researchers to investigate the ways in which the material of buildings and the objects around us is transformed in many ways, far after the time of their conception. In this work, presented on the occasion of the XIIth Venice Architecture Biennale in 2010, their aim was to 'assum[e] a multitude of diverse and heterogeneous types of wear'.[30] This led them to explore very diverse spaces (public places, factories, domestic spaces) and to interview a large number of people responsible for their maintenance, cleaning and supervision. The resulting work

(*Usus/usures. États des lieux/How Things Stand*) is composed of numerous photographs and a series of descriptive and analytical texts. It offers a fascinating foray into the unsuspected world of material deterioration. Its richness is largely due to the systematic inventory it offers of the modalities of material deterioration and the specific problems that its treatment raises, from prevention to detection and repair. Abrasion, scratching, erosion, sedimentation, perforation, fatigue, deformation, dislocation, unravelling and chemical reactions: we discover the wide variety of forms of wear, and the authors show that they are always the results of specific combinations and particular situations.

The interest of Rotor's research, however, does not lie in the creation of a simple bestiary of deterioration. Nourished by a technical awareness of tribology (the science of material wear) and a knowledge of operational maintenance, the approach helps above all to sharpen the eyes of its readers. From the bistro table to the computer keyboard, via the joints of bathroom tiles and the locomotive axle, each case encourages us to immerse ourselves in the ordinary life of materials and to read in their transformations the clues of the relationships formed between objects and their users. This approach, rooted in the world of architecture and design, is explicitly political. Bringing to the fore the omnipresence and variety of deteriorations in matter amounts to 'reassigning to materiality whatever attempts to escape it, revealing what is hidden behind the illusion of the perfect prism'.[31] Architecture and design are pervaded by the temptation of aesthetic perfection. Modernism and minimalism represent the most accomplished forms of this trend, which consists of producing consumer goods and buildings with sleek shapes and perfect lines. The omnipresence of glass, the obsession with transparency or whiteness: these artefacts are designed to cast an abstract material aesthetic, one that denies the very idea of wear and tear. Steve Jackson made a similar plea a few years ago in the field of computer design, encouraging designers to break away from the idea of the 'bright and shiny object'[32] and to pay more attention in their practices to the problems of material deterioration, breakdowns and breakages.

In order to explain the political issues at stake in their action, the members of Rotor use the vocabulary of international relations. They use warlike terms to characterize the posture of most architects and designers in the face of wear and tear. The existence of smooth, 'perfect' objects is in fact based on a fierce fight against decay, an annihilation of all forms of deterioration. However, the whole purpose of Rotor's investigation is to demonstrate that the idea of getting rid of deterioration in this way is a myth, an untenable promise – a deception, indeed, which masks the processes at work and refuses to take into account material life in its simplest expression. Instead of warlike terms, Rotor proposes those of consideration, negotiation and compromise; the members of the collective encourage us to cultivate a 'material diplomacy' in which 'wear is accepted and the difficult questions it poses worked on'.[33]

Rotor's proposal is bold and we think that its possible repercussions, both conceptual and operational, are crucial. In a certain way, it amounts to repositioning and articulating, from the point of view of the mundane life in the Global North where everyday objects abound, two other diplomatic projects developed in recent years in the face of the urgency of the environmental crisis: the creation of a 'Parliament of Things', as conceived by Bruno Latour,[34] and the establishment of a 'diplomatic cohabitation with the living world' set out by Baptiste Morizot.[35] Latour's aim is to escape from the ruts of fixed and exclusive models of science and politics in order to find ways to bring what he calls 'non-humans' into democratic debates. This requires the establishment of new spokespersons (for forests, rivers, etc.) who would not be limited to the register of indisputable scientific facts, but whose hybrid character, both scientific and political, would be fully accepted. For Morizot, the diplomatic quest is played out less in the arena of public forums than in direct contact with the countless beings who share human living environments. It is this common life that matters and requires the reinventing of in situ forms of relationship and communication that are better able to make it again possible to live in shared spaces: in this way, the existence of one species does not require the extermination of another.

If we are here highlighting the work of the Rotor collective, this is because we think it outlines another diplomacy, one that is grounded in the everyday lives of modernity. Addressing the issues raised by both the Parliament of Things and cohabitation with non-human living beings, the point is to pave the way for a diplomacy of *living things* – of things insofar as they are made of living matter, including the least dignified and most commonplace objects. This diplomacy, of course, in no way replaces the political models of Latour and Morizot. But it proposes an additional issue to be placed on the map of political concerns, an issue rarely acknowledged in contemporary so-called 'ecological' debates: namely the relationships between humans and the multitude of things thanks to which they inhabit the world. Bringing to the fore material fragility and the diplomatic questions it raises here means bringing the 'missing mass' of ordinary objects (as well as large infrastructures and the most complex machines) into the domain of ecology. Largely absent from environmental issues, these objects are nevertheless there. They have long made up our milieus in the ecological sense of the term. If we want to be able to debate their roles in our lives and invent forms of use that make sense from an environmental perspective, we must stop taking for granted the view that attributes to most objects qualities of stability, solidity and sustainability while hardly ever putting these qualities to the test. We must find ways to invent a diplomacy that takes material fragility as its starting point.

To this end – and this is how Rotor's proposals continue – we must succeed in bringing the world of design closer to that of maintenance. Those responsible for maintenance, supervision and cleaning do already develop skills of negotiation with wear and tear, skills that lie at the very heart of their activities. They are both the main practitioners of this diplomacy and its best spokespersons. To understand how to start from fragility, to learn how to *deal with* it, we must be able, from the moment of design onwards, to recognize these skills, this situated knowledge, and to draw inspiration from them. In their own way, thus, the members of Rotor encourage us to follow in the footsteps of the maintainers, whom they very aptly portray as 'invisible [...] researchers' immersed in the 'laboratory of reality'.[36]

Care and things

In order to fully explore the consequences of the diplomatic reversal that consists in making fragility the very foundation of our relationships with objects, we must return to the title that we chose to give to the present book and insist on the question of *care*. By highlighting material fragility, maintenance requires us to grasp the political and moral significance of care, understood in all the depth of the meanings gradually deposited in this term in the course of some very productive philosophical and political debates.

There is obviously a somewhat provocative aspect in this rapprochement. How might the notion of care be deployed in an investigation into the maintenance of objects? We are not here claiming that one can mechanically, and without effort of translation, extend the uses of the term 'care' to our exchanges with things, obviously. However, we think that a deeper understanding of maintenance, one that takes seriously the lessons we can learn from those who practise it, involves some of the issues that have been debated in connection with the care of people.

Theories of care – or rather the huge range of discussions that have looked into the question in philosophy, anthropology, psychology and even political science – have developed mainly in the feminist circles of the academic and activist worlds, notably in a series of reactions to Carol Gilligan's work on the links between ethics and the status of women.[37] Let's stick, for the moment, to what these discussions have highlighted regarding questions of vulnerability and fragility. In her work on diabetes management policies in the Netherlands, Annemarie Mol has paved a potentially fruitful path.[38] She distinguishes two 'logics' which organize patient care, first showing that a logic of 'choice' has gradually been established in the Dutch health system, one that aims to give each patient a say in the decisions that concern them. While this policy, which directly involves patients, goes far beyond the previous framework which left all decisions in the sole hands of the medical authority, Mol underlines its limits, particularly in terms of responsibility and mental load. Instead of this logic of 'choice', she proposes a logic of 'care' that in some ways reconfigures the relationship between patients, their bodies, their acquaintances, their loved ones and medical staff in the broad sense. One of the specificities of this logic, explains Mol, is the place it gives to illness and more generally to fragility. Unlike strictly medical perspectives, and even the logic of individual 'choice' – a logic that understands illness as a deviation from the norm, a provisional state of exception that must be remedied – 'the logic of care takes as its starting point our physical embodiment and the fragility of life'.[39] This is clearly, in different terms, the same distinction that the members of Rotor make between war and diplomacy: the logic of care identified by Mol is above all a daily, practical art of *dealing with* illness; a sort of diplomacy, in other words.

This question lies at the heart of the reflections of Joan Tronto, who has sought to identify the contours of the political and moral consequences of care.[40] One of her arguments calls into question the ideal of personal autonomy, so dear to the philosophy of the Enlightenment and still omnipresent in the treatment of sick, disabled or simply elderly people... While there are of course positive aspects to autonomy, it tends to present the need for care as a sign of weakness, a breach in the ideal of individual independence. Care, on the contrary, is entirely focused on the recognition of the interdependencies which characterize human life, and condition the existence of life itself. In a landmark definitional exercise, Berenice Fisher and Joan Tronto described how care practices fall within a very broad domain of activity which includes 'everything that we do to maintain, continue, and repair our "world" so that we can live in it as well as possible. That world includes our bodies, our selves, and our environment, all of which we seek to interweave in a complex, life-sustaining web.'[41] It is obvious from this definition that maintenance activities and their relationship to material fragility are directly linked to care. And the criticism of sleek, forever-new, objects, such as buildings with perfect shapes, is akin to calling into question the ideal of individual autonomy. Starting from fragility as a common condition of people on the one hand, and of things on the other, care and maintenance each affirms in its own way that nothing stands or lasts 'by itself'. People, like things, always need to be made to exist 'a little more'.[42]

Beyond these practical proximities, what is the point of looking at maintenance activities in the light of theories of care? We can identify three avenues that this rapprochement opens up. First of all, the debates around care deal directly with morality and ethics. To draw a link with maintenance is to emphasize that consideration for wear, deterioration and all the activities which consist of taking care of things is a matter of ethics. But, and this is the second fundamental contribution of research into care, this ethics does not consist of a series of principles which would make it possible to define once and for all, in abstract terms, what we are dealing with. On the contrary, the ethics of care is essentially practical. It is an ethics of situation, an ethics which cannot be based on the traditional separation between morality and politics, between ideas and actions. Care is an activity. Care is work. It is impossible to understand it independently of the very concrete operations which it involves.

The same goes for maintenance, in all its forms. It has to be grasped in the great variety of its practical achievements, and it is through contact with these that we can understand its moral significance. Finally, those authors who have worked on the question emphasize the ambiguities of the care relationship and the asymmetries in which care may participate in establishing or cultivating. Most of them thus highlight a third aspect that is very relevant to our investigation into the care of things: vulnerability, and therefore the need for care, is not in their view a weakness. Furthermore, the care situation itself is never unidirectional. This is one of the key points of Tronto's analysis: those who 'receive' care are in no way passive beings. By accepting the attentions of the caregiver, by responding to their activities, by giving signs of comfort or discomfort, they play a full part in care, which must always be understood as a relationship in the strong sense of the term. This point is quite simple to grasp, if only intuitively, in connection with the care that involves human beings. It is less intuitive when it comes to things. However, it is far from negligible. Understanding maintenance activities in terms of care requires getting as close as possible to the relationships that those who carry them out have with matter. But it also means agreeing to investigate the role that things themselves play.

3

Attention

Wear and tear, the deterioration and alteration of everyday objects, are not obvious. We cannot take it for granted that material fragility is the common condition of the existence of things, unless that fragility actually becomes manifest. Experiencing fragility is a matter of perception. People must know how to pay particular attention to things, learn to be sensitive to their variations if they want to extricate themselves from the modern condition which consigns all traces of physical deterioration to the background of most of their activities. By recounting our own experience following in the footsteps of the maintainers of Paris metro signage and by depicting our sudden discovery of the material precariousness of the signboards, we have so far minimized this aspect. However, it is essential to the care of things. In maintenance situations, fragility lies at the centre of an attentional work whose subtleties need to be examined.

Many of the occupations involved in maintenance organize the attention to fragility around one very specific operation: the inspection round. While this undoubtedly does not bring together all the forms of attention involved in the care of things, it nevertheless provides a privileged point of view on the very particular activity of *paying attention*. This will become clearer if we start by describing four rounds of inspection in very different contexts. The first was carried out in 2019 by Étienne, an agent in a public water authority in a town in the south of France, and involved visiting a municipal water tank. The second took place in the early 1990s on the premises of an organization whose photocopiers were maintained by the company that employs Tom. The third inspection took place in 2015 at the Musée du Quai Branly. Here, we follow Lucille, head of the conservation and restoration department, Colas, in charge of preventive conservation, and Lucien, a technician from a pest control company. Finally, the fourth round will take us back to the corridors of the Paris metro at the end of the 2000s, where we will accompany Nadine in the early hours of the morning, as she walks through the spaces of the Quai de la Gare metro station before opening it up to the public. These four scenes will give us a first glimpse of the attentional work through which material fragility manifests itself. We will return to it throughout this chapter, as we seek to understand what this work involves and what it teaches us about caring for things.

Étienne parks his car at the end of a small path on the mountainside.[1] He joins two municipal agents who are busy around a buried structure of which we can only see the roof: the water tank that he has come to inspect. The two agents have emptied it and are cleaning it. Étienne has scheduled his visit in order to take advantage of this operation and avoid having to interrupt the work again. After greeting the agents, he begins his visit by observing the buildings from the outside. The site includes a technical room, a valve room and the tank. Armed with his phone, he takes photos and dictates his remarks into a recording app. There's a displaced tile here, a crack there, a damp spot there. Once this first round has been completed, he enters the technical room and observes the imposing machine which occupies almost all the space. He goes to the valve room, where he draws a diagram of the piping on an app on his phone. He leans silently over one of the valves, examining a joint. He then descends into the tank, stopping on the first rungs of the ladder to observe the interior wall. Arriving at the bottom, he scans the ground with the beam of his flashlight and checks that the slab does not have any steps or unevenness that could cause him to fall during his visit. He then methodically inspects the walls from top to bottom, and has another look at the platform and the pipes. He regularly takes photos and records comments on his phone. Once he has gone back up, he takes the time to chat with the agents who have finished their cleaning session. He asks them a few questions about the history of the site, and the different kinds of work carried out in recent years; he takes occasional notes throughout the conversation. Then, returning to his car, he heads back to the department headquarters located in the suburbs of the main town of the community.

Now a change of scenery: we're in the corridor of a building similar to thousands of others in businesses, public administrations and universities.[2] The long corridor with offices on either side leads to the meeting room. It has a space specially designed for the collective use of a printer and a photocopier. Tom is a technician in the company responsible for maintaining these machines, which are essential to the lives of the offices and employees. During his visit, he was informed of a problem which was occurring more and more frequently with the photocopier – an increasing number of 'paper jams', an annoying problem familiar to every user, where one or more pages get stuck in the depths of the mechanism and interrupt the operation of the machine. He turns on the photocopier. Before looking into the jam issue and while waiting for the machine to 'warm up', he goes through the routine inspection steps, pretty much the same ones he's done on previous machines he's encountered here. Once the cover is opened, he takes a look at the ink circuit and checks whether the mechanisms have become clogged. When the machine is ready, it makes a first copy, which immediately generates a paper jam. He begins to examine, one by one, the different elements of the mechanism: the suction device, the relay chassis, the cycle control switches. He spends a long time looking at a chain and the clutch that controls its movement, then he examines the drive system, looking for signs of wear on the teeth of its gears.

Let's leave Tom in front of the photocopier for the moment, while he prepares to launch a new print, and let's move to Paris, to the Quai Branly museum.[3] Lucien is a technician, specialized in the fight against insects. It's a Monday, the day the museum is closed to the public, and he's accompanying Lucille and Colas on a very special inspection round, which it is hard to imagine could take place in a museum: all three are setting off in search of the fauna that populate the spaces of the institution and coexist with the works on display. If the presence of insects is the subject of their attention, it is because it constitutes one of the main symptoms of the fragility of these works, which are part of the non-European ethnography collections. The Quai Branly houses a large quantity of pieces made from materials that are highly appreciated by many insects who like to take refuge there; some of their larvae use them for their daily meals. This is particularly the case for jewellery and masks. These creatures may be tiny for the most part, but they nevertheless cause significant damage to the plant and animal fibres that make up the works on display. The round made by Lucille, Colas and Lucien allows them to take stock of the population of insects present. It aims in particular to identify the different species, to evaluate their quantities, to understand their reproductive cycles and to document the eating habits of each species. Equipped with a cart to collect specimens and a computer on which they record the immediately identifiable species, the three protagonists make their way through the exhibition spaces without looking at the works on display. Their inspection focuses on the trapdoors that provide access to the fifty-four pheromone traps hidden in the ground. Each time they open one, they carefully observe the places immersed in darkness using their headlamps, note the insects present, gather the corpses of the larger ones and take samples from the traps. Then they move on to the next trapdoor, methodically, exploring this subterranean backstage of the current exhibition.

Finally, let's return to the Paris metro. On this particular day in 2007, it's 5 a.m. and the network will be opening its doors to the public in thirty minutes. Among the agents already on the job, Nadine, an operator at the Quai de la Gare metro station on line 6, is responsible for the state of the so-called 'passenger' spaces, that is to say the areas used by the metro users who pass through on a daily basis. Every day, she carries out at least two inspections (one before opening to the public, the other during the day) in order to identify possible anomalies and report them to the various maintenance departments concerned. With a form and a pen in hand, she walks to one of the station entrances and begins her journey towards one of the two platforms. As she strides along, her gaze shifts from one piece of equipment to the next. Fire extinguishers, lifts, escalators, fixed steps, service gates, metal gates, lighting, exit doors, telephones at the end of each platform, administrative posters, metro maps, display screens, signage modules: she checks everything as she passes, and also comes across a few homeless people who have spent the night here – she noticed their presence thanks to a strong smell of urine which will also be the subject of a report. Her inspection finished, Nadine heads back towards the gate of one of the entrances, where there are already partygoers staggering home to bed, as well as early morning workers, some of whom are undoubtedly themselves maintenance agents who, once they reach their destination, will start their day with an inspection.

What do these brief scenes tell us? What do they teach us about the attention being cultivated here? If we want to take the question seriously and not stick to a superficial observation that contents itself with the piling up of a few synonyms of the verb 'to look', we need to linger on these rounds and describe their specific dynamics, including their most trivial and routine aspects. To begin with, let's consider one very simple aspect: in each case, the maintenance workers we encountered are on the move. More precisely, in all four situations, attention is both the *result* and the *cause* of a change of place.

Displacements

In order to inspect the things they care for, the maintainers we have described go out to meet them. They get close up to them. This suggests one initial way into describing the attention at work here, beyond the somewhat loose vocabulary of observation. Attention seems to be a matter of *proximity*. To be honest, this proximity does not systematically imply a change of place on the part of the people responsible for care. In other situations, it is things that come to them, as with every time we entrust our car to a mechanic so that he can 'service' it, a harmless term which, like that of inspection, clearly expresses the importance of the time and space devoted to attention. In any case, it is indeed a confrontation that is involved. It is in the contact between the body of things and the bodies of maintainers that sensitivity to the subtle material variations of deterioration is activated. The triviality of this aspect should not minimize its significance, particularly because it constitutes an organizing principle for many maintenance activities. Whether it involves setting up a workshop to which objects can be transported (including the most bulky ones, such as train carriages or planes), or ensuring that maintainers can go 'on site' (which then raises the question of accessibility), the organizational assemblage of proximity sometimes represents a real challenge.

This is evidenced by several recent controversies about various situations which have each revealed a lack of maintainers' proximity. Among the growing debates on infrastructure decay in Europe and North America, the question of the decline in resources devoted to the supervision of infrastructure has become particularly sensitive. It is raised again after each serious accident: if the bridge has collapsed, if the gas pipe has exploded, if electricity brownouts have become widespread, this is because these infrastructures have not been sufficiently monitored. In our terms, we can summarize some of these criticisms and concerns in the form of a question: do we devote enough time and human resources to organizing a recurring proximity between things and the people who should maintain them, to ensure that the fragility of infrastructures receives all the attention it deserves? In the late show devoted to infrastructure that we mentioned in the first chapter, John Oliver presented several examples aimed at highlighting the very scarce resources devoted to inspections in many sectors, including the extreme case of hydraulic dams in Alabama, which at that time no longer had any monitoring team. He also mentioned an anecdote which perfectly illustrates the crucial nature of proximity. One day, two entrepreneurs stopped to buy sandwiches in a cabin set up under a Philadelphia bridge. While they were eating, they discovered a large, extremely worrying crack in one of the bridge's piers. They notified the City services, which confirmed their diagnosis, closed the bridge to traffic and began emergency work. The bridge, and with it some of the motorists who used it, therefore owed their safety to a desire for sandwiches, quips Oliver. And, one might add, they also owed it to the lucky encounter between the experienced glances of the entrepreneurs and the cracked pier.

In his book on rust, Jonathan Waldman demonstrates how little this proximity can be taken for granted by describing an important moment in the history of the Statue of Liberty.[4] Throughout the 1960s and 1970s, even though it was visited by thousands of people daily, the statue had actually experienced a form of solitude. Without a full-time manager for more than twelve years, it had apparently not been closely monitored over a much longer period. It was only during one very particular episode that things changed and that the problematic nature of the absence of any proximity with maintainers was revealed in its full extent. On 10 May 1980, two activists began to climb the statue to hang a banner demanding the release of Geronimo Pratt, a member of the Black Panthers convicted of murder eight years earlier. Although it was barely twenty-four hours before they came back down and surrendered to the police, the operation had considerable repercussions. It led to renewed attention to the actual material of the monument.

Worried about what these activists were capable of – as nobody knew anything about them – David Moffit, the first permanent manager of the site after a long period of vacancy, had spent the day tracking their movements using his binoculars. As he scrutinized the surface of the statue, he ended up seeing dark spots that seemed to be holes. Convinced that the activists had used pitons for their climbing, he was horrified. But he was not reassured when, on their arrival back down, he saw that they were in fact only equipped with suction cups and claimed not to have damaged the statue in any way. Taking another look through his binoculars, he discovered that the spots he had seen were in fact far more numerous than he had expected, and could be seen across the entire surface. His fears were confirmed, after he sent one of his team's members to look inside the statue. The spots were indeed holes – hundreds of holes that had nothing to do with the climbers. The statue was being eaten away by corrosion. And no one, until then, had noticed it.

This short sequence of events (lasting barely two days) led to a long and hazardous process of restoration, from asking the advice of the top international experts to carrying out the work itself, including the difficult matter of raising public and private funds, which led to the statue being closed to visitors for several months and ended with the celebrations of its centenary in July 1986. Waldman's account of this vast campaign is fascinating, but it is the first movements triggered by the presence of the activist climbers that matter to us here. By illegally coming into contact with the monument, they made its surface an object of unprecedented concern. They *drew the attention* of the site manager to it and allowed him to take a fresh look at the statue, which he had never approached in this way before, and which he had therefore never seen from this angle. First by using his binoculars, then by going there with his team, he completely changed his point of view of the interior and exterior of the monument, discovering the variety of materials that composed it, and becoming attentive to the actions of the rust which had spread to such an extent that it threatened the integrity of the entire statue. In other words, the sudden arrival of the climbers created the conditions for an unprecedented proximity between the maintainers and the Statue of Liberty, a proximity that reorganized their attention and brought to the forefront of their concerns a material fragility which, until then, had remained unsuspected, buried in the apparent immutability of the monument and its imposing presence.

To understand the kind of attention the care of things draws on, it is therefore important to first recognize the role of contiguity, which shows that maintenance is a matter of co-presence, of keeping close guard, of 'staying with', as Steve Jackson puts it.[5] The story of the rust discovered by chance on the Statue of Liberty, like that of the crack in the bridge pier that was fortunately located next to a sandwich bar, shows that the absence of conditions arranged to cultivate this proximity – the latter being of any value only when it is constantly reiterated – means insufficient concern for material fragility.

However, we must not fall into an overly reductive vision of the idea of rapprochement which could suggest that immediate presence and the contact of bodies are the only sources of an attentive gaze. Certainly, the four scenes from which we started underline the importance of physical proximity, but they should not lead us to discredit forms of remote inspection. Just as the person responsible for the Statue of Liberty first approached the surface of the monument using binoculars, there are many situations where the configuration of attentional proximity plays out far beyond physical co-presence. In her book *L'Engagement dans le travail* (*Commitment at Work*), Alexandra Bidet shows how the operators of a telecommunications company manage to compose different forms of proximity with the network despite the distance involved, by cultivating 'contact' with the dynamics of communications traffic on the supervision screens, which they browse and 'inhabit' as worlds in their own right.[6]

By focusing solely on the importance of the physical proximity of maintainers and the object of their care, we also risk missing another essential dimension. If attentional work were solely determined by proximity, it could be summed up by the organization of a face-to-face meeting at which each protagonist would remain immobile. Why not, after all? Perhaps one way of being attentive could consist of remaining impassive, of erasing the traces of one's presence to allow oneself to be imbued with the signs of fragility that one would end up seeing as they detached themselves from the surface of things, after an adequate time of silence and calm. But this seems doubtful. In any case, this is not at all what the activities of the agents with whom we opened this chapter suggest; quite the contrary – they encourage us to think that attention is also deployed in *movement* itself.

This second dimension of the displacements involved in maintenance can be summarized in the very idea of a round of inspection. The attention cultivated by those who carry out such an inspection is in fact organized by, and in, the process of walking around. Étienne walks around the site he is visiting before moving into the tank, and once he is inside, continues to survey the platform; Tom goes down the corridors from one photocopier to another; Lucille, Colas and Lucien follow the location of the traps where they collect their insect samples; and Nadine walks up and down the metro station for which she is responsible. Again, these trips may seem quite trivial. However, they reveal an important aspect of the capacity of each person to be sensitive to the fragility of the objects, machines and infrastructures they are in charge of. Far from being just a consequence of the requirement for proximity, walking around is an attentional operator in its own right. This idea lies at the centre of the ecological theory of perception developed by James J. Gibson.[7] His work breaks with dualistic conceptions, which separate not only the actions of the body from the skills of the mind, but also human beings from the world around them. Grasping the ecological dimension of perception amounts to bringing together the presence in the world of beings (including animals) and the matter of things which present themselves to sight as much as they are seen. It is the dynamic encounter of bodies and things that generates perception in Gibson, not the attitude of a being endowed with a mind capable of predefining the relevant features of the world.[8] In this dynamic, movement is essential. It is through movement that matter sticks out and the promptings of things are manifested (Gibson calls them 'affordances').

Following Gibson we can understand that the perception of fragility is not the result of a meeting played out in advance between material properties on the one hand and the mental dispositions of the humans involved in maintenance on the other. It is a situated process, through which both the features of fragility and the attention of those who examine things by constantly varying their point of view gradually emerge. This goes well beyond the scope of the inspection itself, during which the main signs of fragility that need to be taken into consideration can be identified. By carefully observing the activity of maintainers, we also discover more subtle movements: a few steps forward or backward, a body leaning forward, a head tilting to one side then back. Standing over the photocopier he is examining, Tom spends his time changing position, each time modifying the scope of his observations. Nadine pauses several times during her journey to approach an object, look sideways at a wall or turn her head at a fork in the metro corridors. Whether it is a matter of getting a little closer or, on the contrary, taking a step back, paying attention to fragility clearly requires not only establishing the right distance between maintainers and things, but also a continuous adjustment of this distance, as if the success of perception did not consist in establishing adequate sensory parameters, but on the contrary in continually modifying the focal length.

In the previous chapter, we insisted on the difficulties that the external observer may encounter in spotting the signs of the fragility of things. By following movements and adjustments of distance, we realize that the attention in question is partly played out in forms of perception, *agencements* that differ from usual situations of use. If maintainers can see material elements that others do not see, this is not, or not only, because they have developed 'purely' sensory capacities, sharper than ordinary. It is more specifically because they *configure their sensitivity* to suit the situation. Varying their point of view through movements of greater or lesser magnitude, they establish a dynamic relationship with the things they need to take care of – a dynamic that allows fragility to manifest itself.

Multisensoriality

In many situations, however, the gaze, even when moving, is not enough. In the spaces of the metro station that she inspects, Nadine shows herself to be very sensitive to odours (smoke, urine, animal corpses, etc.), some of which must be reported, while others, not necessarily any more pleasant, are considered the 'normal' scents of the place. The work of Tom, the photocopier repairman, regularly relies on his listening to the mechanisms of the photocopier to identify possible signs of damage or malfunction.[9] What comes next, in the sequence described by Julian Orr, clearly shows this. Tom repeatedly presses the button that triggers the copies without ever looking at the result of the operation, but listening carefully to each noise generated by the machine. In this way, he identifies the various elements engaged during the process: the motor that starts, the chain that moves the mechanism, the internal switch that locks, the clutch that releases... For each of these elements, he tries to identify by ear either its defective activity (because it is weakened), or its effective action (but one that has no effect on the printing), which could be the cause of the paper jam. During his investigations, he focuses more specifically on two aspects. On the one hand, the characteristic sound produced by the mechanism which starts moving at each new step constitutes a valuable indication for identifying a possible discrepancy with that produced when the machine is operating correctly. On the other hand, the precise moment at which each element is triggered in the printing cycle of a copy forms another clue in the quest to locate the problem. Throughout his exploration into these sounds, he develops possible scenarios and reduces the scope of his subsequent actions.

This is another essential dimension of the proximity of bodies: it provides the possibility of a *multisensory contact* which nourishes attention well beyond the dynamic focusing of the gaze. But while hearing and smell hold an important place in certain areas, the sense most used after sight is, unsurprisingly enough, touch. Placing their hands on things, sliding them behind fixed elements in order to reach invisible parts, touching surfaces with their fingertips, activating more or less resistant mechanisms... These are all tactile operations carried out to appreciate the properties of the materials at play, operations that seem essential to the manifestation of fragility. Let's go back to Étienne as he enters the tank. After carrying out a first series of checks, he makes a gesture which surprises the observer accompanying him. He passes his hand over the interior wall of the structure, rubbing the surface. He then rolls his fingers together and shines his flashlight on them. From this gesture, he establishes a first observation about the quality of the cement which was used in the construction of the structure, which he considers mediocre given the sand that stays on his skin on contact with the wall. This sand crumbles a little too easily, suggesting that the wall is porous. While this first assessment is not enough to make a firm diagnosis of its condition, it still guides the rest of the visit. Étienne says he will need to be particularly vigilant when inspecting the adjoining valve room and ensure that no trace of seepage is visible there. And even if the visit to the room in question does not reveal anything suspicious this time, the observation made by simply passing his hand over the wall is translated by Étienne into the notes which he dictates to his smartphone. It will then be entered into the database which collects his reports. During the next visit, between twelve and eighteen months after this, the condition of the wall will be the subject of even greater attention, undoubtedly initiated, yet again, by direct contact between Étienne's hand and the surface of the wall.

As Maria Puig de la Bellacasa points out in *Matters of Care*,[10] touch is unique in that it is a symmetrical sense. The person who touches is simultaneously touched by the object or person with whom he or she comes into contact. This transaction adds an important aspect to the focusing of attention. It enriches the meaning of the term 'proximity', which we have been so far using without defining it. When they touch things, the people who take care of those things initiate a relationship that is a matter of intimacy. Puig de la Bellacasa explains that this intimacy is a form of intensification; it exacerbates concerns, engaging those who carry it out in an exchange from which it is impossible to escape. If it would be vain to compare the senses term by term, gauging their greater or lesser capacity to involve people (you have only to think of the use of violent music broadcast at very high volume when someone is subjected to torture, or the truly unbearable nature of certain odours), it is important to underline the amplification of attention at work in the tactile relationships of maintenance – especially since touch is omnipresent in the care of things and sometimes intervenes in situations where we might not imagine that it could play a role.

To get a better idea of this, let's change the scene again. We are now in the streets of Paris and following the work of the employee of a company mandated by the city council to remove graffiti from the city's facades. Since the early 2000s, graffiti removal has been organized in Paris as an urban maintenance activity. Like many other cities around the world, the council has been inspired by the policies successively put in place in New York in the 1980s and has set up a way of tackling graffiti which no longer aims at the definitive eradication of the practice, but on the contrary apprehends it as a recurring, almost natural phenomenon, requiring constant attention and daily interventions.[11] Thus, for more than twenty years, several companies have been responsible – following a multi-year call for tenders – for detecting the presence of graffiti in the capital on a daily basis and removing it.

It's a weekday in Paris, in March 2017. It's 7:42 a.m. José parks his truck on rue de Cléry, in the 2nd arrondissement. Once the engine of his vehicle is turned off, he grabs the form summarizing the work he needs to carry out today in this street. Each line of the table designates a graffito to be removed, specifies the date of detection, the main characteristics of the surface on which the graffiti have been made and the estimate of the size of their surface expressed in square metres. José raises his head several times in the direction of the facades and locates each of the unwanted inscriptions noted on the form. He then gets out of the truck, approaches the first graffito indicated on the table, examines it for a few seconds, then touches it with his fingertips. Riveted on the coloured letters, his eyes accompany his hands, which make light movements along the length and then over the height of the graffito while varying the pressure. The gestures are precise and meticulous. They are also fleeting – barely a few seconds. They are routine: we have already seen José, like his colleagues, perform them on several occasions. From an external point of view, this short caress may look like a ritual – like a greeting that an executioner addresses to his future victim, with his outstretched hand and his frank gaze, before drawing his weapons to slay him. Of course, the scenarios have little in common. This gesture, José explains to us, is a way of finding out about the situation. It helps him to gauge two types of material properties: those of the graffito and those of the facade. By touching, he appreciates the consistency of the paint, its granularity or, in other cases, the thickness and nature of the ink. He thus gauges its ability to resist the products available to remove it. At the same time, he assesses the density and porosity of the material on which the graffito has been inscribed, and this allows him to get an idea of its greater or lesser tendency to absorb paint or ink. He also estimates the capacity of this material to withstand the attacks of different removal techniques.

Physical contact is thus a key element in the rest of the operation. While they are assessing the different materials they are dealing with, the graffiti removers identify the methods they will need to deploy in order to carry out the removal: they ponder which product, which instruments, will provide the optimum combination for making the graffiti traces disappear as best as possible while preserving the properties of the surface on which it was affixed. This is what José explains to us as he walks towards some graffiti done in ink: 'On shop windows, there's sometimes an opaque plastic film, and it's important to check if it's placed on the indoors or outdoors. If it's outside, then in removing the graffiti we'd damage the plastic film – a disaster!' These few words express how essential is the ability to appreciate the subtleties of the materials involved in this kind of maintenance activity.

Although they know what shop windows or marble facades are potentially made of, such as stone or concrete, and even if they are able to guess what most of the graffiti they approach has been applied with, the graffiti removers we have observed confirm that each intervention requires a tactile exploration allowing them not only to verify and validate their first observations, but also to discover, in the play of urban textures, the roughness invisible to the naked eye, the material idiosyncrasies of the alloy formed by graffiti and the surfaces. With touch, they develop a 'new sensitivity to the surface of things',[12] thanks to which they take stock of the material fragility of this portion of urban reality that they are responsible for maintaining.

During his research with auto mechanics, Tim Dant became very interested in the tactile aspect of maintenance.[13] He showed that the contacts of the material with the skin in work situations, generally via the hand, fuel a 'sensual knowledge', one that it is not easy to represent outside of bodily experience. To illustrate this point, he describes the actions of a technician trying to understand how the damaged door of a van that has been involved in a traffic accident behaves. During this sequence, although many of his movements involve his hands and sometimes his entire body, none are intended to modify the door in question. He does not exert force or pressure to straighten it, nor does he manipulate any tools. He simply operates the door several times, focusing on the opening and closing moments. Like Tom, the technician observed by Orr, he regularly changes points of view, going from the outside to the inside of the vehicle. He also varies the speed, sometimes opening the door slowly, sometimes slamming it suddenly. At each stage of this operation, which lasts around twenty minutes and includes thirty-five openings and closings, the mechanic touches the handle, the surface of the door in several places, its hinges, and the surrounding bodywork. Though apparently innocuous, and marked by the monotony of their repetition, these gestures are essential, explains Dant. First, it is through them that the mechanic understands the condition of the door, precisely identifies the elements that are damaged, and hypothesizes about the conditions that have led to this damage. Second, as with graffiti removers, these gestures guide him in the preparation of the rest of his intervention, projecting from these points of contact the methods to be put in place and the instruments for repairing the door.

If we limited ourselves to the gaze alone, we could be content to believe that attention to fragility is mainly a matter of recognizing previously identified superficial signs. Even by insisting, as Gibson does, on the ecological nature of visual perception, we would miss the very specific action of touch through which maintainers seek to 'make matters speak', as Johan M. Sanne puts it in his discussion of the work of railway maintenance workers in Sweden.[14] More than an encounter, touch establishes an exchange during which maintainers become sensitive to the singularity of the heterogeneous assemblages of which things and their environments are made. Sight, hearing and smell of course play a role in the apprehension of certain aspects of these assemblages, but touch is the main operator of attention to their mutations. By caressing facades, by plunging their hands into the grease of a machine's mechanisms, by feeling the resistance of a moving part with their shoulder, by assessing the heat of an electrical component, maintainers are like the 'artisan', whom Gilles Deleuze and Félix Guattari present as 'one who is determined in such a way as to follow a flow of matter'.[15] They connect to the tiny variations that punctuate the life of the things they care for.[16]

There is therefore nothing transparent or passive in the attention cultivated by maintenance. Nor is it the work of a purely intellectual attitude or a moral position which designates in advance a series of salient characteristics to be systematically taken into consideration. Attention to fragility is at once situated, active and open. The sensitivity that develops here is the result of more or less sophisticated *agencements* which provide matter itself with the possibility of manifesting itself just as much as they adjust the always embodied perceptions of those who take care of things.

Expertise

We suspected that the sensitivity at work in maintenance situations is a matter of material expertise when we realized that most maintainers were capable of identifying many features in things that we did not know how to perceive, and we can understand it even better now that we have looked at the attentional techniques they perform. In the type of consideration it gives to things, this expertise recalls that which crystallized among art specialists during the second half of the nineteenth century, on the basis of which the historian Carlo Ginzburg conceptualized a specific form of knowledge which he described as a model based on the interpretation of clues.[17] Ginzburg's demonstration is well known. He explains how a method of attributing paintings developed by Giovanni Morelli in the nineteenth century inspired not only the creation of the character of Sherlock Holmes by Arthur Conan Doyle, but also the principles of psychoanalysis defined by Sigmund Freud. He puts this moment into perspective by demonstrating that if it seems to mark a break in the different fields concerned, it is in fact part of the lineage of a form of ancestral semiotic knowledge which, from the art of hunters to graphology and medicine, developed in parallel with, then at the margins of, the model of the modern sciences based on mathematics and the highlighting of regularities.

Ginzburg's analysis is of particular interest to us because it places the question of attention at the centre of the 'clue paradigm'. The latter is in fact based on the ability to turn away from the most obvious features of the object studied so as to concentrate on details usually considered insignificant. For Morelli, this attentive gesture amounts to completely overturning traditional methods of attributing pictorial works. Rather than focusing on the aspects commonly associated with each painter and established as characteristic signs of their particular school, precisely those aspects that forgers strive to reproduce as faithfully as possible, he argues that the expert needs to observe more anecdotal elements such as fingers, nails or even earlobes. This is where copyists lose focus and make mistakes. And, contrary to what was the general view among the profession at his time, Morelli states that it is also in these forms, apparently without great value, that the personality of the painters is most clearly revealed. It is also a question of ears in the example that Ginzburg takes from Conan Doyle. In the short story entitled 'The Adventure of the Cardboard Box', published in 1892, Sherlock Holmes manages to deduce an unexpected kinship by first carefully examining the victim's severed ears, then by scrutinizing those of Miss Cushing. We recognize in this capacity for deduction inaccessible to ordinary mortals the trademark of the detective, who never ceases to surprise Doctor Watson, his faithful companion, by focusing on the periphery of the crime scene, by discovering signs where no one else sees them. Like Morelli, Holmes focuses on small things to resolve the enigma posed by each situation and to reconstruct the truth: tiny anatomical particularities, footprints in the mud, the ash of a cigarette… And like them, Freud focuses on details of speech to reveal part of the mysteries of the unconscious: innocuous words, stammers and, of course, the 'slips of the tongue' that fascinated him.

The attentional *agencements* of maintenance are organized on very similar principles. Rather than resulting from an exhaustive overall view focusing on the structural aspects of the thing to be maintained, the manifestation of fragility involves the same concern for details. As he tours the site where the water tank he is inspecting is located, Étienne keeps spotting barely perceptible traces, which it is difficult to imagine relate to the structure he is responsible for maintaining: the state of the vegetation, a difference in the colour of the cover… This is also the case for Nadine, who does not stick to the most obvious elements of the metro station she is in charge of, but dwells on details that are insignificant in the eyes of external observers: the safety sealing of fire extinguishers, the surface of the signboards, the condition of the steps, the material of the tactile strips, the level of brightness in each zone… Her experience of the station has little to do with that of the vast majority of travellers who pass through it, whether they are occasional visitors or long-time regulars. This discrepancy is also obvious in Tom, the photocopier repairer. Orr writes that his inspection starts at the periphery of the machine itself, by the nearest waste-paper bin. There, like a seasoned Sherlock Holmes, Tom looks for the presence of pages thrown away after the latest 'jams'. When he finds some, he observes them and can read valuable clues in the crinkles of the paper, thanks to which he gains a first idea of the photocopier's behaviour and plans the rest of his intervention. And what about the way in which Lucille, Colas and Lucien approach the spaces of the Musée du Quai Branly, on the lookout for the tiny world of insects? Not only do they find traces of pests which must at all costs be prevented from multiplying, if some of the works on display are not to disappear, but they also follow the trail of relatively harmless animals (like spiders) which inform them about the state of the museum's atmosphere: whether there is too much dust, or other problematic insects, for example.

Like Morelli, Holmes and Freud, maintainers pay attention to what is usually overlooked in order to find tell-tale signs in matter. They act like connoisseurs. Specific to the clue paradigm, this relationship to things relates to a form of expertise that is quite different from the scientific posture that seeks to identify universal laws and insists on the virtues of abstraction. Ginzburg explains that attention to traces and symptoms endeavours to grasp each phenomenon in its singularity. It is an essential aspect of the care of things and the attention that nourishes it. Far from dealing with inert objects, reducible to a series of properties fully known in advance that just need to be monitored mechanically, maintainers treat things individually. They dwell on their specificities and understand them as unique, even when they seem completely standardized. This is obvious for the spaces and works of the Quai Branly museum, but it is also true of the other cases that we have mentioned. It is not any old tank that Étienne inspects, nor an ordinary metro station that Nadine supervises. They are very particular structures, about which both already have a certain amount of information that guides their attention, and that is supplemented with additional elements after each visit. Likewise, Tom knows each of the photocopiers he is responsible for in all their idiosyncrasy and treats them in accordance with their history. He knows for example that the machine we have seen him inspect is already old and that some parts are worn out. He also knows that it is no longer used at a sustained rate. This has consequences for the propensity of some of the components to become clogged more quickly than on other machines. He also keeps up to date a file on each of the machines he takes care of, detailing the interventions carried out since the start of his maintenance, as well as the list of parts replaced or repaired. Similarly, graffiti removers adapt their activity according to what they have gradually learned about certain neighbourhoods, certain sections of street, even certain facades. They know which ones react badly to a particular type of treatment, the dosage that is needed and the degree to which weather conditions need to be taken into account, even if the material on which the graffiti have been made ought to be able to withstand the latter. They also identify areas where a particular graffiti writer becomes more active, and adapt the rhythm of their interventions in order to temper their enthusiasm as best as possible.

As Ginzburg points out in another article, the type of knowledge at work in such attention to details and singularities pertains to a form of intuition that is 'rooted in the senses'.[18] It is an intuition that human beings are all capable of, and that they even share with the animal world. This clarification is important, since for Ginzburg it is a question of emphasizing the gaps between the cognitive processes which take place in the clue paradigm and the pretensions to rigour and exhaustiveness found in the scientific model inherited from Galileo. Relying on traces on the periphery of things, connoisseurs, doctors and detectives, as well as maintainers, accept an element of irreducible uncertainty that cannot be swept under the carpet of statistical regularities. Each and every one carries out methods that are 'eminently qualitative' and have 'as objects individual cases, situations, and documents, *inasmuch as they are individual*.'[19] However, as Ginzburg points out, the term 'intuition' is 'tricky' and we must be wary of it. This is particularly true in our case, since it could encourage us to see an attention to fragility as based solely on the sensory skills of maintainers and the tiny gestures they perform on the material. This would be a mistake. A large number of maintenance activities are based on hybrid forms of attention combining sensory skills and more or less sophisticated instruments.

At the Quai Branly museum, some of the insects that need to be recorded cannot be seen with the naked eye. The samples that Lucille, Colas and Lucien take have to be examined under a microscope. In this same museum, the conservation department works in close collaboration with technicians who develop monitoring instruments allowing the interior of certain works of art to be observed in order to develop maintenance methods that do not further weaken them. These tool-assisted inspections are also common in the industrial world, where numerous techniques are used to assess the condition of certain parts of a machine to a degree of precision much greater than that of a traditional 'manual' examination. This is particularly the case for so-called 'non-destructive testing' methods, implemented to carry out a diagnosis of the internal structure of an object without risk to its physical integrity. Radiography, magnetic scanning, tomography, ultrasound exploration: the technologies are varied and remind us that in certain cases the fragility to which we need to pay attention escapes the sensory capacities of humans. The proximity that is established in this type of attentional scrutiny then takes on a completely different dimension. Thus equipped, maintainers can go beyond the boundaries of bodies and cross surfaces to locate traces of cracks, flow or crumbling, which are all processes that would signal a point of no return if they extended so far as to become perceptible to humans.

The use of these instruments is obviously reserved for very specific situations, and the issues this raises should not weigh too much on the way we apprehend the care of things – something that we seek to do in as broad a way as possible. However, it does tell us about one important aspect of attention to fragility: its relational nature. Depending on the objects concerned, the context of their use and the maintenance situations themselves, any trace of a material mutation can be considered sometimes as negligible, sometimes as critical. If those who take care of things teach us to see material fragility as a condition common to all things, the heterogeneity of their practices, something that we have only touched on so far, reminds us that the degree of attention that we are prepared to pay to this fragility, the scale on which we understand its manifestations and the technical resources that we will devote to it vary greatly. It is never fragility 'in general' that emerges from the attentional gesture, but a specific and singular fragility, configured in an attentional gesture that sometimes involves a significant use of tools.

We will gain a clearer understanding of this aspect of maintenance if we consider another type of equipment for attention, one that is much less spectacular but much more widespread. Let's return to our initial four scenes. None of their protagonists maintains an 'immediate' relationship with matter. Everyone has a valuable instrument at hand that they use in each of their inspections. In the metro station, Nadine circulates with a printed document of several pages on which she writes a few notes and regularly ticks the relevant boxes. This shows that her inspection is structured around certain predefined aspects, such as the presence or absence of certain kinds of equipment in the station, all listed, along with their visual appearance and their operating state. Likewise, Étienne's examination of the water tank involves regularly referring to an observation sheet, the different items of which he has drawn up with the manager of his department and which he will complete from his audio recordings once he is back in his office. Tom also takes forms with him when he checks the photocopiers. And the survey of insect traps at the Quai Branly museum carried out by Lucille, Colas and Lucien goes hand in hand with the filling in of tables and computer files. In their case, the inspection is also governed by the international standards of Integrated

Pest Management and its bio-infestation indicators which must also be reported after each round.

Whether they take the form of loose pages, notebooks, mobile-phone apps or computer files, these pre-structured templates are common to all maintenance occupations. They are essential to the collective dimension of the work, since they enable the results of inspections to be circulated within the services concerned.[20] It is thanks to them that it is possible to prioritize and plan interventions, but also to monitor the things maintained over time, whose progressive changes in state can be appreciated by comparing the archived records. Each scene that we have described is thus part of a network of activities based on the information collected on site using these documents, and then entered into dedicated databases or 'computerized maintenance management systems' (CMMS). But these documents also play an important role in the actual completion of the inspection and thus the organization of attention. In the files, lists and other forms, a series of points are assembled which comprise an outline of the thing to be maintained. Among all the possible ways of understanding it, this outline represents a very specific version of this thing that has reached its shape through historically and organizationally situated processes to which maintainers must be sensitive. Many of the 'details' that guide inspections are contained in this outline. They do not simply represent the main sites for identifying traces of fragility: as with Morelli, Holmes or Freud, these details also have an ontological value. They are distinctive features.

Since they designate 'what matters' in the thing to be maintained and are linked to more or less precise indicators of fragility, the documents that travel with maintainers guide their sensory explorations. Their omnipresence in inspections highlights a process that goes beyond just the occupational situations on which this chapter has focused: attention to fragility is always selective. It is deployed in close relation with a definition of the thing whose maintenance it helps to ensure. This definition is not systematically formalized or stabilized once and for all, it is not always detailed, it is – of course – never exhaustive, but it is inseparable from the attentional gesture, whether it is carried out in a professional context or in a more ordinary situation. Paying attention to the material fragility of an object, whatever its size or nature, always comes down to selecting from the depth of its physical presence certain traits that are set up as essential characteristics as and when one becomes sensitive to them. In this sense, attention is generative. It is the first moment of a maintenance that is part of the very existence of *things* that need constantly to be brought into being, rather than *objects* 'already there' that simply need to be reproduced mechanically.

Vigilance

By choosing to explore the sensory aspect of attention to fragility before addressing the expertise demonstrated by maintainers, and then highlighting the role of the instruments they use, we have endeavoured to shake off the reflexes of certain social sciences which favour an institutional or organizational reading of the professional practices of maintenance. While there is no question here of denying the role of the rules and norms which underlie interventions – quite the contrary – we believe that it is absolutely crucial to take into consideration what is at stake in the situation, in the encounter between women or men and the things they take care of. This is not only so as to emphasize the inevitable adjustments to which these rules are subject, as in all work activities, but above all to discover how attention is also very often deployed beyond the scope of formalized prescripts.

An approach based on the practical depth of maintenance activities in particular prevents the care of things in occupational situations from being reduced to the transparent and routine implementation of pre-established protocols. It is tempting, in fact, to see the lists, tables and forms that accompany certain maintainers as self-sufficient guides that simply need to be followed literally in order to take material fragility into account. Attention, from this point of view, is akin to a control operation carried out on the basis of stable criteria with respect to which a state of conformity or non-compliance needs to be established. We have seen that this is far from being the case. The documents in question do not exhaust the attentional gesture, even if they help to configure it. And this gesture is not reducible to a fully controlled gaze that imposes itself on an inert and passive matter. However selective it may be, attention to fragility is anything but a disposition closed in on itself, inscribed solely in normative criteria and incarnated in particular bodies by means of an occupational apprenticeship. Because it takes shape in contact with things and their singularities, it connects the focus on definite features to a more floating sensibility, one that is open to the unknown. By carefully observing certain areas, by scratching surfaces, by touching walls, by knocking on materials to make them resonate, those who practise maintenance become sensitive to the variations which are expressed in the very folds of matter.

This active and open part of attention is difficult to observe. It rarely provides an occasion for detailed explanations from the people concerned, who have generally cultivated it on the job, developing their skills through these gestures whose value lies in their incessant repetition. Sometimes it is more obvious. When we accompanied Étienne on the visit to the site where the water tank that he was to inspect was located, we witnessed a scene which confirmed that, despite the importance of the indicators guiding his inspection, indicators that were all the more relevant since he himself had helped to define them, his attentional work also consisted of making himself alert to the manifestation of unexpected signs.

The following episode took place after he had finished his visit. Eager to collect as much information as possible in order to populate the database in which all the structures for which his department is responsible are listed, he had been talking for a few minutes with the maintenance workers cleaning the tank. One of them was leaning against the exterior wall of the valve room. A few seconds after he had taken his back off the wall, Étienne stopped in the middle of a sentence, stared at him and said: 'Wait a bit, what's that noise? Do it again, just to see...' The worker in question, half amused, half worried, repositioned himself back against the wall, before moving away from it once again. A loud 'plop' sounded. We all looked at Étienne, who was already approaching the wall. He began to touch the surface of the wall, pressing hard with his hand, sometimes managing to reproduce the suspicious noise. He commented out loud: 'I can't believe it. What on earth's going on here? The paint's completely peeled off.' We could indeed see his hand sinking slightly in places, revealing the presence of paint blisters invisible to the naked eye. After continuing his tactile and visual exploration, he turned around to explain: 'The colours are a bit different, there, right up high. I'd never have guessed. This may mean that there's a seepage issue, or just that the paint wasn't properly applied. It's outside, so actually it's not very serious. We need to discuss it at headquarters.' He started taking photos with his phone, as we had seen him do during his inspection. But the photos, meant to provide an intelligible trace of the problem that he could transmit to his colleagues, didn't satisfy him: 'I'm going to film it, in fact, it's the only way to make them understand.' Placing his hand back on the wall, he finally recorded some videos of the movement of the paint blisters.

This scene contrasted with the routine that had until then prevailed throughout his visit. We can also guess that it's rare and that not each inspection gives rise to a surprise of the same kind. Nevertheless, it teaches us a lot about the open nature of the attention to fragility. The care that Étienne brings to the site is reflected not only in the series of sensitive probings that he makes to compare the state of the materials present with the pre-established criteria detailed in his files, but also in the free-floating attention that he deploys, even when the inspection sequence itself is completed. Without a specific goal or a determined focus, this form of 'low-intensity' monitoring leaves wide open the field of possible variations and the forms in which fragility can become manifest.

Francis Chateauraynaud has described this dynamic connecting the routine of stabilized indicators with an openness to the unexpected by drawing on the notion of 'vigilance'.[21] Based on three apparently contrasting situations (being in charge of the control room of a nuclear power plant, air navigation and ordinary car driving), he shows that the very unfolding of the action involves sometimes highly detailed formal rules coexisting with in situ improvisations nourished by largely tacit sensory explorations. It is the dynamic relationship of these two registers of activity that characterizes what he calls vigilance. One of the main challenges here is to enrich the understanding of work activities, and in particular to move away from the hyper-formalist framework specific to so-called risk industries, which focuses simply on an ever-expanding list of rules and protocols and defines the skills of operators on the sole basis of verifying ever more precise and numerous criteria. The quality of vigilance lies in the delicate balance between 'relaxed perception and cognitive tension'. In other words, we must recognize the value of the permanent adjustments that are at work in vigilance and, as in the phenomenology of Maurice Merleau-Ponty, grant all the senses their place in the exercise of perception.[22] Even if it relates to specific situations, this analysis is valuable for us, since it rehabilitates 'open' sensory action and thus underlines the wealth of workers' skills over and above their sole ability to follow rules. Most of all, it does not deny the importance of these: indeed, it is the coexistence of the two forms of engagement in action which nourishes vigilance.

But it is when he draws on the philosophies of Paul Ricoeur and Gilles Deleuze to describe vigilance as an operator of responsibility and a form of presence to the world that Chateauraynaud helps us to specify more precisely what is at stake in attention to fragility. By broadening the spectrum of vigilance beyond contexts of proven risk, we can in fact concur with him that this attention consists, in those who cultivate it, in 'assuming [their] participation in the course of things'.[23] Being attentive to fragility means engaging in an uncertain inquiry which involves an exchange between the bodies of humans and those of things. This exchange is not obvious in ordinary use, and the attentional gesture is aimed at making it possible: this consists precisely in making room for things 'themselves', for their own ways of behaving, of reacting to our approaches and of surprising us, sometimes. If we need to know how to 'make matters speak',[24] we perhaps need to know how to listen to them in the first place, and grant the things we want to take care of an opportunity to express themselves in the relationship we have with them.

Attachments

The people whose work we observed may occupy, for the most part, low-status positions, on the fringes of the main activities of their respective companies, but they act like connoisseurs. They know how to focus on a series of predetermined details that they manage to isolate from among the material abundance of things and their environments; they remain attached to the singularity of each entity they take care of, and are capable of accepting their excesses, of becoming sensitive to their calls. These attentional skills cultivate an 'art of espousing the propensity of things'.[25] They are obviously not innate or simply acquired through theoretical training. They are the fruit of an experience in contact with things and are densified by the continuous repetition of gestures and circulations – the constantly reiterated present which, as we have seen, characterizes maintenance. The *agencement* of attention to fragility is never stabilized once and for all. Constantly guided by previous experiences, it does not completely reinvent itself with each intervention. It is nourished and draws its strength from the tireless renewal of care.

By paying attention as connoisseurs, maintainers also help us get a first idea of the type of knowledge involved in the care of things. Because it is carried out as closely as possible to the matter involved and its tiny variations, maintenance is deployed in a dynamic relationship with knowledge. Not only does it mobilize more or less formal, sometimes extremely sophisticated, types of knowledge about materials and their behaviour, it also creates many opportunities to discover an irreducible part of the life of things in situation. In this sense, taking care of things always also means *getting to know them* ('*faire connaissance avec elles*').[26] And it is this attitude, both curious and respectful, that we would encourage while being attentive to the attention cultivated by those who take care of things.

However, to fully understand what is at stake in the attentional *agencements* of maintenance, we must return to the starting point of our questioning. We began this chapter by claiming that fragility, the material condition to which we seek to make ourselves sensitive, cannot be easily grasped. This is what we highlighted from our own experience in the corridors of the Paris metro before the signage maintainers initiated us. And this is also what we have insisted on throughout our description of the work of the people we have observed over these last pages, emphasizing the richness of their attentional expertise. But what exactly do we mean when we say that material fragility is not easy to grasp?

If we want to answer this question, we must ask three others: what are we talking about? from where are we talking about it? and who exactly are 'we'? The absence of signs of wear, the relative invisibility of marks of fragility are not intrinsic properties of objects that impose themselves everywhere and on everyone. Quite the contrary. The film *Gagarin*, directed by Fanny Liatard and Jérémy Trouilh, illustrated this magnificently by painting the portrait of a young man so devoted to the housing estate of Ivry-sur-Seine he lives in, in the deprived suburbs of Paris, that he undertakes to ensure the maintenance of its buildings, hoping to save them from their planned destruction. Faulty electricity, dilapidated plumbing, seepage, crumbling – the film shows what everyone who lives or has lived in this type of building is very familiar with: an omnipresent, even oppressive fragility, with regard to which there is no need to develop exceptional attention skills. Many geographers and anthropologists have also highlighted the highly visible and omnipresent nature of urban infrastructure fragility in the countries of the Global South.[27] Claiming that material fragility 'is not obvious' is therefore a situated observation that is valid only if it is itself questioned, and ultimately turned back on itself: how is it that in certain places, in the eyes of certain people, the fragility of things is so difficult to see?

What we have here, in fact, is a characteristic feature of the consumer experience which has gradually taken hold in rich countries (or more precisely in certain areas of rich countries) since the beginning of the twentieth century. Relying on objects and infrastructures that are always available, the model of mass consumption has, symmetrically, established the idea of a consumer who does not have to worry about the state of the things he or she uses. We will need to return in much more detail to this form of contemporary consumption, which also leaves a very special place for maintenance itself, and for those to whom maintenance is entrusted.[28] Sticking to the question of material fragility and its invisibility for now, we discover another aspect of the 'ecology of attention' of modernity.[29] The point here is no longer to understand what saturates the attention of a large number of the citizens of rich countries, but to realize what is withheld from it. What is commonly called the 'consumer society' is based on a systematic obliteration of the material fragility of things – a denial of wear and tear that configures the negligence of consumers, delegating the concern for fragility to a small segment of the population placed at the service of the fiction of the solidity and durability of modern artefacts. Just as it is possible to draw inspiration from trackers or certain artists to learn how to rediscover the subtleties of what naturalist 'landscapes' hide and to become sensitive to the living world that surrounds us,[30] we can learn to observe this fringe of the population at work, and find, in the attention that it cultivates, the means to extricate ourselves from the modern regime of the consumption of objects, a consumption that pays so little attention to things.

We have mainly focused here on people whose job is maintenance and who are, in some way, professionals of the attention to fragility. But we can of course find very similar attitudes among others, in particular those called 'amateurs', those men and women who develop a love of things that precisely allows the latter to express themselves; people who want to be surprised and transformed by things in a two-way attachment.[31] In her book on old Ford Mustang owners in Europe,[32] Cornelia Hummel describes with great respect their bonds with the objects of their passion.[33] She shows repeatedly that the attachment to these American cars, which are quite rustic and which have numerous mechanical weaknesses, draws on an intimate relationship cultivated in the actual usage of the car, a contact with matter very close to what we have been describing in this chapter. This is particularly striking when it comes to driving. In Europe, Classic Mustang owners generally come from less privileged backgrounds. Their car and its daily maintenance represent a very significant budget, and each breakdown or accident threatens their ability to continue with the relationship. Driving a Mustang is therefore not a trivial operation. Through the testimonies she has collected, and her own experience as an owner, Hummel highlights the attentional work found among drivers, whose bodies become sensitive to signs of material fragility, sometimes to the point of excess.[34] No sooner have they settled down in their vehicles, once the ignition key is turned, their senses are activated. First comes listening to the engine, as it idles and then while driving. Listening combines the pleasure generated by the purring of the V8 and increased vigilance to the slightest irregular or unusual sound, the symptom of a potential problem. In addition to constantly listening to the engine, they also listen to the noise emitted by the shock absorbers, brakes and gearbox: the owners quickly become experts in these areas. With their hands on the steering wheel, their backs pressed against the seat, their feet on the floor and the pedals, owners are always touching their Mustang. Through this contact, they can gauge the greater or lesser firmness of the steering, its symmetry, the tension of the brake cable, the state of wear of the pads. The rigidity of the rear axle amplifies the impression of being one with the vehicle and thus guides their attention to the trajectory adopted in a bend in the road, even its more or less pronounced bumpiness, of which they sometimes need to be wary. Hearing, touch, sometimes smell, and of course sight: each of these senses is on the alert, connected to the components of the car and its environment so as to identify the slightest clue of a failure that would require intervention before a more serious breakage occurs. This attentive driving, which is based on permanent vigilance, differs greatly, in its perceptual density, from the ordinary driving of a general public vehicle. The attachment to the car and the care given to it entail the near impossibility of avoiding the signs of its fragility. Hummel also says that in certain cases, the situation becomes untenable and that owners must learn to reduce the intensity of their vigilance if they are to continue to enjoy driving their Mustang. If it is essential to be able to listen to what things say about their state, one must also be able to preserve the conditions for their use. It is important that the thing, just because it is given credit for being able to express itself, does not take complete control, thereby completely hindering usage.

There is obviously no question of imagining that 'we', rich citizens of rich countries, children and parents of the 'consumer society', can all, at every moment and with every object that we handle, adopt this form of use, one that is entirely absorbed by the concern for fragility. On the other hand, we can try to develop a certain attachment to some of the so-called consumer goods and the more imposing objects that constitute our living environment, such as buildings or roads – a relationship that draws inspiration from both maintenance workers and Ford Mustang enthusiasts. Perhaps by loving, at least a little, the things we use, we will be able to become attentive to the fragilities that the world we live in today endeavours to hide. This requires the cultivation of an intimate material relationship, alert to the slightest unevenness and aware of potential overflows.

4

Encounters

Tim Dant has spent several years studying the world of automobile garages and observing the activity of the mechanics. At the forefront of research on maintenance, he has developed an original analysis completely out of step with the way in which the social sciences had hitherto approached the car, namely as a consumer good, emblematic of the modern condition or as the main instrument of contemporary mobility. Without denying these dimensions, Dant looked at an aspect of the life of automobiles in which sociologists, geographers, historians and economists had never really been interested: the time these automobiles spend away from roads and car parks, in the hands of the workers who ensure their maintenance, overhaul and repair. Dant went to meet those people (mainly men) who strive to ensure that cars continue to operate in the best possible conditions. As we mentioned in the previous chapter, he particularly endeavoured to highlight the sensory part of the work of mechanics, whose bodies are in almost permanent contact with the vehicles. In his meticulous descriptions, he shows the details of the material encounters that lie at the heart of maintenance. Observing the reassembly of an engine,[1] he shows, for example, how two mechanics adjust their movements with great fluidity, without using plans or measuring instruments. He sets out the circumstances of a sensitive relationship during which the two men first gauge the parts with their eyes, then grasp them one by one, feel them and thus locate their place in the complex assembly of the engine. Each time, specific gestures are performed – a twisting of the wrists, elbows and torso, but also a shaking and even tapping – that allow the alignment of the pieces to be finalized and the operation to continue in a material intimacy that needs few words.

Dant's texts are full of examples which confirm the importance of material exchanges between maintainers and the things maintained. In line with the phenomenology of Maurice Merleau-Ponty, he highlights

THE CARE OF THINGS

the embodied experience of mechanics, whose tasks, as he shows, are always simultaneously perceptive, cognitive and manipulative. It is in this 'pragmatic of material interaction'[2] that knowledge of the things that maintainers take care of is developed, and their experience and expertise refined over time.

While he mainly depicts fluid interactions to illustrate the intimate nature of these exchanges and the virtuosity of certain mechanics, Dant also mentions some slightly less harmonious material encounters. In order to highlight the capacity for improvisation that the operators demonstrate throughout their interventions, he takes, among others, the case of Ray, whom he describes in the process of removing a front wheel. The operation is obviously not new for the mechanic; however, it generates some friction. Despite the tools at his disposal, Ray is not immediately able to loosen the first bolt that holds the wheel to its hub. He then assembles an ad hoc tool from a lifting key to which he adds a long tube to exert additional pressure. He places both hands at the end of this additional handle and, with arms outstretched, presses with all his weight, bending his knees as the key progresses, millimetre by millimetre. After a few seconds of intense effort, the bolt finally gives way and loosens. Ray then removes the extension he had added to his key and continues unscrewing the bolt effortlessly.

The scene is trivial. Amid all the routine garage activities, it could be easily overlooked. But it is because Dant wants to capture the slightest gesture of the mechanics (he photographs and films them several times) that he pays attention to this fleeting moment. This puts us on the trail of an aspect of maintenance that we have minimized until now: sometimes, things resist. Even in the hands of experienced and well-equipped workers, they sometimes don't allow them to do what they want with them. Or not right away, and not completely. Any of our readers who has changed a wheel knows this full well, and has undoubtedly experienced it with much more difficulty than Ray, who had done this task many times.

These situations should not be taken lightly. By beginning our journey with the question of the fragility of things, then following the trail of the attention brought to bear by maintainers, we have so far described material relationships as relatively peaceful. We have insisted on the importance of a form of respectful proximity – on concentrated looks, then touching, scratching, even caressing; we have realized that the art of maintenance is closely linked to this ability to listen to things, allowing them to express themselves in various ways, depending on each situation. Being attentive to the fragility of things requires that we recognize and acknowledge the material action that never ceases to pass through them, and keep pace with the flows of transformation they are made of. Now, if we follow the thread of this argument, we also need to accept that things sometimes do as they please, that one of the ways they express themselves is simply by rebelling. In an inspection, this resistance translates into a form of opacity: an inaccessible crack, signs of wear hidden between two parts, inaudible friction, a slow seepage which makes its way into the concrete for weeks on end without leaving any trace. And during interventions, the indiscipline of things can take more spectacular turns, when they no longer simply evade examination but directly oppose the actions of those who are handling them to ensure their maintenance. Just as human beings sometimes strive to escape the care they receive, things can be recalcitrant in the face of those who seek to maintain them.[3] There's no need to go far to be convinced of this. Let's stay at home and in the surrounding environs: the radiator bleeder seizes up, the hot water tank safety valve refuses to make the quarter turn that the heating plumber recommended be carried out at regular intervals, the main block of the electric shaver can't be opened so we can change the battery, the washing machine filter can no longer be pulled right out, the chain guard prevents the bicycle derailleur from being properly oiled, the roller blind mechanism has got stuck in a place inaccessible to adult hands, the computer declines to install the latest version of the antivirus software, the lawnmower blade refuses to be disassembled, the head of a screw no longer provides any purchase for the screwdriver tip, the edges of the nut have worn down so that the spanner can't turn it...

The examples of recalcitrance are countless, and the reasons for them very different. Some forms of resistance are the simple result of wear and tear, others can be attributed to design flaws or inadequate after-sales servicing, and yet others to inappropriate use. We will return to these aspects in our final chapter. What matters to us at this stage is the way in which these forms of material resistance shift our gaze to the *sensible encounters* involved in maintenance. All of the examples that we have just mentioned could in fact be described from the point of view of the person confronted with them. Each time this happened, we would see rather less tender gestures than those we depicted in the previous chapter: we would witness harsher contacts, we would also probably hear swearing, or even the echo of a few knocks and bangs. These situations show us that, over and above attention alone, maintenance is also a matter of hand-to-hand struggle which can change into rough-and-ready exchanges, sometimes marked by a certain brutality, which contrast with the image of a considerate maintenance entirely dedicated to the preservation of things whose material integrity is untouchable.

Recalcitrance

Let's focus for a moment on the bodies of the women and men involved in maintenance, before returning to the bodies of things. In many ways, activities undertaken to ensure the material longevity of objects can inspire aesthetic trends. It is indeed tempting, once we have evoked images of people in direct contact with artefacts and seeking to prolong their existence, to lapse into a celebration of bodily gestures and an exalted glorification of the physical, immediate relationship that is established here between humans and things. Many maintenance operations, hardly ever noticed, lend themselves to this enterprise. The description then turns into a sensual portrait of an almost romantic bodily encounter. This trend is found in Matthew Crawford's book *Shop Class as Soulcraft*.[4] The book quickly became a classic. It takes the form of a fable. The author meets Fred, a mechanic who teaches him the art of repairing motorcycles. In doing so, he unknowingly provides Crawford with the means to rebuild his life and finally give it meaning. Crawford, who had been working in a think tank, ended up abandoning the intellectual world (until he rediscovered its charms, as shown by the fact that he wrote this book, which was followed by two others) and opening his own repair shop. This confirmed the revelation he had received: it is here, by getting his hands dirty – an expression that we have become accustomed to use when we celebrate the wisdom of these people who can deal with the 'reality' of the material world – that Crawford starts to enjoy work again. It is here, he insists, that the very meaning of work lies, in this face-to-face encounter with a matter that sometimes cooperates and is sometimes recalcitrant, in this physical bond and the ultimately practical kind of intelligence that it mobilizes. And Crawford highlights the emptiness of his past activities, the absurdity of the rules and formats of the intellectual life he had led until then, a desperately abstract and artificial life. There are, of course, many fascinating reflections in this book, and most of them directly mirror our concerns as well as those of Mierle Laderman Ukeles' long-term investigations. One of Crawford's manoeuvres consists, for example, in re-evaluating so-called manual jobs and questioning the systematic primacy given in school careers to the most academic trajectories, which systematically exclude good students from those manual 'trades' or 'occupations'.

Of course, we share this concern, and our approach amounts precisely to making maintenance activities in all their richness really 'matter'. But when Crawford starts to insist a bit too much, we no longer follow him, particularly when he suggests that working in his repair shop is more authentic than the work carried out by executives or academics, and that the advent of information technologies will tend to make manual work disappear. Let's pass over the stigmatization of intellectual work and the clear separation drawn by the author between the office and the workshop. Let's focus instead on his book's central idea – that the repairer's 'real' work tends to disappear with the gradual breakdown of the true (i.e. 'material') link between human and machine. It's an attractive suggestion. But it doesn't stand up to a careful observation of maintenance practices. We have seen that there is no reason to distinguish a priori between the numerous modalities of apprehending matter, whether they involve touch or smell, or instruments which extend and equip sensitivity, and touch and feel in their own way.

If this aesthetic of the authentic relationship with matter proves problematic, it is above all because it can lead to neglecting the dark side of our relationships with things, and to leaving aside – or even glorifying – the violence that sometimes characterizes the close encounters involved in maintenance and doesn't leave the maintainers unscathed. This is precisely what the recalcitrance of things alerts us to: the commonplace nature of the occasional violence of maintenance activities, and the suffering that it can cause among the people who carry them out. Ask your plumber, or take a look at the creaking joints of the technicians at the workshop where you leave your car or bike. Think back to the posture which Tom, the photocopier repairman described by Julian Orr, is forced to adopt.[5] Imagine the muscle pain of the graffiti removers, after just one week of rubbing the surfaces with their rags to remove the obstinate ink, pressing their special scrapers to the stickers to remove them from the street furniture, or handling the spray gun which projects hot water or a mixture of sand and silica onto facades at high pressure, avoiding staying on the same spot for too long. Think about the after-effects caused by repeated more or less easy contact, with the asbestos so often found in the areas reserved for the maintenance of buildings, or with radioactive materials during the maintenance of nuclear power plants.[6] In their recalcitrance, things do damage in return, and sometimes cause real harm. They wear out those who maintain them. A description that would only highlight the positive traits of work as always satisfying and rewarding because it is physical and as close as possible to matter would be misleading and ineffective. Although it might be benevolent, it wouldn't tell us much about maintenance. It would also neglect the fact that, in many situations, maintenance pertains to what sociologist Everett C. Hughes calls 'dirty work'[7] – a thankless, undervalued activity that a small, privileged part of the world's population delegates to a large majority of poor workers living on the margins of an arrogant consumerist capitalism.

On the front line of maintenance, recycling and waste management, this bodily violence is ubiquitous in the countless sites in countries of the Global South where electronic instruments abandoned by consumers in rich countries pile up. A great deal of research has documented the activities carried out in these places.[8] One example: based on interviews and observations with people who collect electronic components for resale in the four main electronic waste markets in Dhaka (Bangladesh), Mohammad Rashidujjaman Rifat and his colleagues provide us with a striking description of the physiological consequences of the labour involved in maintenance, particularly for the hands, which they depict in radically different terms from the caressing and caring hands depicted in our previous chapter.[9] Here, 'hands have become hard, black, often swollen, and with marks of old and new injuries in multiple places'.[10] They suffer daily from getting cut by the sharp edges of metal or glass parts, and from electric shocks in tests carried out without protection: these shocks end up damaging the entire nervous system. The hands of these workers are the preferred entry point for pollution from the lead and cadmium present in various components, but also from the octane found in products used for washing.

For the most part, maintenance is a matter of working-class activities, in the South as in the North, and it does not escape the harshness of the working conditions that have long been described by ergonomists, historians and sociologists of work in their studies of musculoskeletal disorders.[11] On many professional sites, the pleasure of physical engagement, the mastered coordination of gestures, the virtuosity deployed in the handling of instruments, all entail the progressive wear and tear of the body, the pain caused by tirelessly repeated positions, and the bodily deformations specific to certain trades.[12] Instruments bruise, mechanical parts on the assembly line hurt. Metals, dust, solvents, plastics, grease, paint: all kinds of substances penetrate the skin and become embedded. From temporary discomforts to chronic illnesses, including disability, we must never minimize the effects of manual work on the body. But we must also not go to the extreme and seek at all costs to paint an unnecessarily gloomy portrait of maintenance work. The hardness of physical activity and masculinity are combined in mechanical maintenance, and suffering itself is sometimes the driving force behind the meaning of work, notably because it demonstrates what is fine and useful in the experience of a trade often undervalued by others.[13]

Without falling into the caricature of miserabilism, then, we also need to avoid any effect of romanticism that would amount to minimizing the violence that occurs in certain exchanges with matter. If the men and women who take care of things have their own importance, and if we wish to grant them due recognition, this is also because they engage their bodies in the process of maintenance, they intimately associate their bodily movements with those of things, whether they are acting in an occupational setting or not. And while they teach us to become attentive to the incessant transformations of matter through their own attention to it, we must also keep an eye out for the signs of their own wear and tear. This is the other lesson of the sometimes brutal recalcitrance of things, quite apart from any realization of their autonomy in the manifestations of their existence. It teaches us that the sensible encounters of maintenance both unfold in and foster an *ecology of fragilities*. The care of things whose material fragility is recognized goes with the commitment of bodies that are themselves fragile, and very often weakened by the very conditions in which they are put to work.

THE CARE OF THINGS

Disassembly

Let's continue this line of thought. Scenes of material recalcitrance also remind us of one aspect that should be obvious but may seem counterintuitive to those who believe that the care of things is just the expression of a concern solely focused on physical integrity – as if maintaining mainly amounted to ensuring that the objects in question remain intact. The cases that we have described until now clearly show that this vision is too simplistic. More conflictual material encounters reveal situations where maintenance, far from remaining on the surface of fixed things, involves disassembly. Whether it is cleaning inaccessible areas, adding materials (to lubricate mechanisms, for example), or even repairing or replacing individual parts: when you want to take care of certain objects, you sometimes have to take them apart, disperse the components, before assembling them again, putting the whole thing back together – a process that then needs to be carried out again and again. This, of course, is a truism for most workers. And the automobile garage is the ideal place to realize this dimension of maintenance. You just have to poke your head into a workshop to see that, without being completely 'taken apart', the cars are far from remaining intact during the mechanics' interventions. Actually, dismantling proves to be essential for maintenance in very varied contexts, from the upkeep of weapons – which, of course, is based on the incessant reiteration of their disassembly and reassembly – to the management of urban networks that, over a much longer time, sometimes involves replacing entire sections.[14]

Lara Houston's observations of mobile-phone repair shop owners in Kampala, the capital of Uganda,[15] show clearly the extent to which phones are dismantled in the process of repairing them. Her striking photos show worktops littered with open cases stacked on top of each other, scattered electrical and electronic components and cables of all kinds. To a lesser extent, this is also what we can observe in the ethnographic drawings of Anaïs Bloch and the descriptions of Nicolas Nova in *Dr. Smartphone*.[16] Such places, far from the ordinary situations in which these technical objects are used, are the scene of real operations of dismemberment and decomposition; they remind us that, apart from wear and tear, and breakdowns, maintenance activities themselves can give objects a hard time.

The first time we took stock of the actions involved in disassembly dates back to a specific episode in our study of metro signage maintainers. That day, we followed two agents who had to take care of a sign called a 'PLT' for *panneau lumineux transversal* ('transverse light panel'). As its name suggests, this type of signboard is hung across a corridor (and not along a wall) and has its own lighting system. The intervention in question consisted of replacing the PVC sheet slipped into a metal box, displaying obsolete information (one of the metro lines concerned had recently changed terminus). Until then, as we said above, we had mainly been struck by the attentive skills of the maintainers, from their detailed knowledge of the principles underlying the entire signage system to their abilities to identify fragilities that were far from obvious. During this operation, we discovered a side of their activity quite unknown to us.[17] What we thought would be a fairly simple replacement task actually turned out to be a much longer and more delicate procedure, which began when one of the agents, perched at the top of a large stepladder, half-opened the signboard housing whose backlight was still on. He started by sliding his hands inside, then part of his head, to work out how to remove the PVC sheet, while avoiding burning himself on the neon. After several attempts, punctuated by shaking, grunting and swearing, we saw him take out a large metal frame in which the said sheet was housed. But this wasn't the end of the matter. Once the signboard was placed vertically against the wall of the metro station, we observed the agents struggling on each side of the frame with the numerous small screws (sixteen in total). Hearing the two men cursing, it was difficult for us to know if these screws were there to ensure the box was watertight and keep the main sheet in place, or quite simply to deliberately hinder the action of the maintainers. It was only once all the screws had been removed that they were able to remove the front face of the box, remove the sheet of PVC and replace it with the one they had brought. They still needed several minutes to carry out the entire operation in reverse and complete the intervention.

Disassembling and reassembling things to make them last: part of maintenance is much more dynamic than one might think at first glance. What these situations also underline is the very relative importance of the ideas of integrity and uniqueness in the care of things. This echoes a point we made in the previous chapter about paying attention

THE CARE OF THINGS

to fragility. Part of maintenance is accomplished in a relationship with objects that could be called 'disaggregation', a relationship which undoes their apparent physical unity to deal with components or materials that are temporarily removed from a 'totality' whose importance is thereby placed in apparent abeyance. We saw that this focus on details fostered a form of expert inquiry, essential to the adoption of an attentional position. By detaching themselves from an overall visual, auditory and olfactory apprehension which grasps objects as whole entities, maintainers become sensitized to the material mutations at play. In this way they can concentrate on portions, sometimes tiny elements, that seem insignificant to their habitual users. Situations of disassembly extend this logic of disintegration, and themselves offer opportunities to continue, or even radicalize, the investigations of those who take care of things. Digging through the components, checking the condition of parts that are difficult to see, observing unsuspected mechanisms: dismantling an object helps to discover facets of it that were previously inaccessible. Maintenance then becomes exploration.

Blanca Callén and Tomás Sánchez Criado have described the subtleties of this material inquiry, based on the investigations they carried out in Spain with two types of actors involved in the recovery of electronic objects: an informal group of immigrants who collect waste in the streets of Barcelona and two collectives based in Madrid, specialized in the repair of electronic devices.[18] These waste pickers rely on fairly similar operations, despite the obvious differences in their activities. Each seeks in their own way to make the objects they collect last: they set up 'vulnerability tests'[19] aimed at identifying what can actually be maintained from among the multitude of materials gathered. Most of these tests are based on dismantling operations that are not always simple, but in which the salvagers have gradually become experts, and thanks to which they investigate the material states of the objects concerned, their composition and their functioning. A computer found in the street thus finds itself caught up in a series of experiments, from a simple power-up confirming its potential reactivation to tests to see if the electronic circuits and the various components will work (motherboards, graphics cards, memory sticks and hard drives, to mention only the most obvious). These operations gradually disassemble then reassemble the computer, while defective elements are repaired or replaced.

These situations are in many respects exceptional, and we cannot of course generalize them by claiming that each disassembly carried out during maintenance amounts to completely dismantling the object concerned, or that it gives rise to a material investigation in-depth. However, they have the merit of highlighting certain important dimensions of maintenance encounters. We've already mentioned the first: some objects are easier to disassemble than others. The case of telephones and computers perfectly illustrates the range of possibilities in the field. If the repairers in Kampala and Madrid, like the recyclers in Barcelona, manage to dismember certain machines until they have piles of components that they can reuse or replace, this is because they were manufactured in a modular manner and all their elements meet specific standards. The possibility of their dismantling has therefore been integrated into the very design of these machines. This is obviously far from being the case for all objects. As we have seen, a certain number of them resist dismantling for various reasons, and exploration is then only possible at the cost of engaging in struggles whose results are unpredictable.

But these cases can teach us something else about disassembly and maintenance. Of the objects whose parts we find scattered in repairers' workshops, some will never be reassembled. Their dismantling is not intended to ensure their own continuity. Like scrapped cars, these objects have become reservoirs of spare parts which allow the lives of other telephones, other computers (or other washing machines, other printers, other solar panels, other CD players, etc.) to be extended. Like dead trees continuing to nourish the forest that surrounds them, they gradually disintegrate by donating their components to things that can still be made to last. This metaphor, of course, has its limits, since not all of the elements that make up these objects will end up merging into others, and many residues will need to be subjected to other treatments, while still others will be significant sources of pollution. But the process at work is nonetheless very important for understanding how broad the scope of certain maintenance activities is, despite their apparent modesty and the banality of the objects they concern. In the contemporary capitalist regime, artefacts that no longer function are considered waste. At best, they are decontaminated, and some of their materials use recycling networks that, as we now know, fall short of the promises of the circular economy.[20] On these sites,

THE CARE OF THINGS

generally set up on the fringes of Western consumption circuits, they are diverted from their fate as waste in order to play their parts in another history of technology and consumption.[21] A history in which supposedly obsolete devices continue to be functional, and those that are considered definitively unusable have their various components and materials fed into the maintenance of the former.

Much more than a simple operational issue, disassembly is therefore a matter of material ethics and politics extending from the meeting between the body of the thing maintained and the bodies of its maintainers to the setting in motion of flows of materials which extend encounters well beyond the singular object. Exploring the places where these dimensions are particularly evident encourages us to fully appreciate what maintenance does to things: the successive disassembly and reassembly operations involved are not trivial and do not leave things unscathed. When it comes to disassembly, the sometimes dizzying question of the transformative aspect of maintenance arises.

Transformations

Let's return for a moment to the Kampala repair shops studied by Lara Houston. In addition to the material flows and 'cannibalization' processes in which certain terminals feed on the remains of other phones in order to survive, it's the inventiveness shown by repairers that stands out. They deploy part-replacement techniques to free themselves from certain incompatibilities, which – as they eventually demonstrate – are quite relative. Unplugging, displacing and re-soldering, they adjust, modify and retool the various devices. One striking practice in the eyes of Western observers accustomed to seeing new or barely damaged terminals consists in setting up bypasses on the motherboard circuit and installing new connections. While these connections permanently render certain features unusable, they allow the phone to perform what repairers and their customers consider to be its main mission: making telephone calls.

This type of operation differs radically from other forms of maintenance that we have encountered so far. It reveals in particular an aspect to which we will return in detail in the next chapter: in certain cases, taking care of things involves significant functional and formal modifications, fully accepted by those who carry them out. Maintaining something can sometimes mean transforming it significantly.

By following the trajectory of a water pump in use in Zimbabwe, Marianne de Laet and Annemarie Mol examined this apparent paradox, showing that the technology in question had managed to last not because it had demonstrated foolproof robustness, but on the contrary because it had been able to transform itself according to varied types of usage.[22] In contrast to the idea of an all-conquering innovation that owes its success to its hermetic and immutable character, as suggested by the sociology and history of science, the Zimbabwe pump has allowed itself to be appropriated by its users, who have ensured its maintenance by means of many transformations. One adjustment made by members of the different villages in which a model of this manual water pump was being used involved a well-cut section of an old tyre, which sometimes did the trick by replacing the initial leather seal. While the large bolts assembling the main block to the lever were supposed to be regularly tightened, some pumps operated without any bolts, as steel bars had been inserted into the holes to ensure the assembly held in place. These modifications did not in any way represent circumventions of what was considered normal use. On the contrary, they had been made possible by the design of the instrument, which had been guided by a concern for simplicity and ease of maintenance. Most parts could simply be replaced and readjusted with tools available in local communities, and the pump's instructions themselves detailed very flexible maintenance procedures.

In their article, De Laet and Mol use the term 'fluid technology' to characterize this type of object, capable of lasting not *despite* a series of transformations, but *with* them, *thanks* to them. In doing so, they seek to mark their difference from technologies whose design principles aim to avoid as much as possible the modifications made by users and the adjustments that repairs in situ generally require.[23] But if we shift the discussion to maintenance practices as such, we can ask ourselves whether it is really necessary to isolate a particular type of object. In fact, this could be counterproductive and suggest that certain objects might be expected to last without undergoing any transformation. Now, while obviously not all maintenance interventions assume such radical forms as those seen in mobile-phone repair workshops in Kampala or around

Zimbabwean wells, it is important to recognize that no form of care leaves things completely unscathed.

To fully understand this, all we have to do is turn our attention to objects which at first glance seem fixed and immutable, but which – as we discover by observing them through the eyes and the hands of the maintainers – are also constantly changing. This is particularly the case for buildings of all kinds structuring the urban landscape, which can appear as immobile entities, literally indefeasible. Architectural masterpieces in particular play fully on this effect of permanence and immutability through their minimal and transparent aesthetic.[24] But once we are willing to observe them in real use, we realize that buildings continue to evolve over the years. They are subject to adjustments, deterioration, adaptations, repairs and sometimes even radical reorientations.[25] Their life after design and manufacture is rich in twists and turns which come into dissonance with the sleek and abstract vision that architects and urban planners had created in their stories and polished images. What is true over the long term is just as true from the point of view of daily life. If we take the time to follow the activity of caretakers and other technicians on a day-to-day basis, we realize to what extent a building, far from being a mass of inert matter, 'lives' and evolves in small changes and repeated modifications.[26] Radiator valves are replaced, seals changed, windows repaired, cracks filled, air-conditioning systems adjusted, roofs renovated, walls repainted...

We can also broaden the picture and consider larger urban settings. This is what Rob Shaw did, accompanying a team of night-time street cleaners in part of Newcastle upon Tyne.[27] He shows how essential these maintenance workers are to the life of this neighbourhood, where executives and employees pass by during office hours and partygoers arrive at the start of the evening. By following their movements amid the young and festive crowd, their repetitive gestures and the instruments they handle, we realize what their cleaning activity does to this city of two faces. Sweeping, picking up and gathering, the agents gradually transform disparate elements of the bustling nightlife into refuse that they remove from the public spaces of the neighbourhood. Flyers and advertising leaflets, fizzy-drink and alcohol bottles, cigarette butts and food wrappers accumulate, then pass from the state of consumer goods to that of waste, following the movements of the cleaning machines. The list of material transformations that allow the night-time city of leisure to once again become the daytime city of white-collar workers does not end there. Broken glass and streaks of alcohol and grease from various forms of take-away food leave marks that the agents do their best to remove. Cleaning up bodily fluids is also crucial. The repeated presence in the same corners of the city of urine and vomit, with their corrosive properties, represents a real threat to the surface of facades. In addition, there is of course the weather situation. Rain, cold and heat cause various reactions that make the treatment and concealment of these materials more or less difficult. In this management of the flow of materials generated by nightlife, the city is not simply cleaned, in the sense of being cleared of undesirable elements and returned to its previous state. Maintenance operations circulate heterogeneous materials that they gradually change into waste by adding water and different maintenance products, which they agglomerate and then redirect straight into the sewers or towards the waste treatment circuits. They help transform the city in the same way as the festive activities which leave traces of all kinds. Neither the streets nor the facades emerge completely unchanged from this daily merry-go-round.

It is exactly the same with the graffiti removers in Paris, whose practices we have already mentioned. Despite the vocabulary sometimes used in political speeches or calls for tenders, erasure is irreducible to mere removal and never returns the graffitied surfaces to their state 'as was'. Whatever the technique used, removal involves transformation. Splashing water or silica on stone, like using chemicals on wood, windows and metal, affects surfaces. These substances also produce residues, most of which are discharged into the sewers, while others remain for a while on the pavements or road surfaces. Even more blatantly, erasure very often takes the form of masking. The graffito is removed by applying a layer of paint which covers it and adds to the materials present. Just like the nocturnal operations of Newcastle's cleaning agents, each erasure therefore acts directly on the material texture of the city.

We saw in the previous chapter that those who are involved in maintenance demonstrate a sensitivity to the material changes in things. Their attention to fragility goes with a sensory knowledge of the constant transformations that affect what they take care of. However, the different cases that we have mentioned so far suggest that all maintenance is, in a more or less blatant way, transformative. Rather than an external action that would aim to fight head-on the mutations identified upstream of each intervention, the care of things is more akin to fostering and playing a part in their becoming. It actively produces a form of 'sameness', whose subtleties we still need to understand.

The attentional gesture therefore does not mark the beginnings of an interruption – quite the contrary. It even extends well beyond the time of inspection and contributes to the transformative interactions of maintenance. This is because the more or less subtle mutations at play in the care of things cannot extend to infinity. There is no room for relativism in the world of maintenance: not all transformations are acceptable. There is always a risk of losing the thing which is taken care of, seeing it being so transformed that it disappears despite maintenance – or worse: because of maintenance.

Worries

As acts of care sometimes marked by a violence that leads to all kinds of transformations (of the thing itself, of its environment, and also of the maintainers), the sensible encounters of maintenance are haunted by the risk of disappearance or too radical an alteration. They are never completely mindless and cannot be carried out lightly if they are to succeed.

As we have just mentioned, the seemingly banal and routine operation of removing graffiti on city walls offers a fine illustration of this menace. Let us return to the precise description of the gestures it involves. In the previous chapter, we stuck to the visual and tactile examination of surfaces. We discovered that removers take the time to caress the facades they are dealing with in order to become sensitive to the fragility of the materials present. However, this movement is not only used to establish a diagnosis of vulnerability. It also guides the conditions of removal as such by facilitating the choice of the technique to use. Depending on the characteristics of the surfaces, certain removal methods are delicate matters, while others are simply prohibited, since their use would seriously damage the facade in question. In each 'manual of specific technical clauses' – included in the successive calls for tenders published by the City of Paris when renewing service contracts – one can find a table displaying 'prohibited pairs of method/surface'. Reading them, we learn that polished stone should not be spray-cleaned with water, or that the application of chemicals to a plaster with minerals is prohibited. This table of simple rules helps avoid disasters. However, it is not enough to make removal a trivial action. Once the appropriate technique has been chosen, removers proceed with caution.[28]

To understand the subtleties that this involves, let's get together again with José, whom we encountered in the previous chapter in front of a store window. After running his fingers several times over the graffitied surface, he confirms that no plastic film is present on the outside of the window in question. The use of chemical solvents is therefore possible. But José caresses the window a little longer for another reason. As he explains to us, the ink is sometimes mixed with acid which eats away at the glass. The result is that, while it is possible to remove the ink, the inscription sometimes remains partly visible, or even clearly legible. Despite a momentary hesitation, he decides to go for it. He returns to his vehicle and returns with a canister marked 'VG Graff', a green scraper and clean rags. He pours a small quantity of pink, grainy product onto his scraper, spreads it on the window in small circles and lets the product act on the inscription for a short time. Over the course of the circular movements of the scraper, applied without pressure, the letters become deformed, the ink liquefies and everything turns into a dark blue paste. José seems reassured: 'Actually, it's pretty good! I can feel the ink coming off and it's not rough underneath.' (He runs his gloved fingers over the surface.) 'I can feel it's staying smooth.' He continues making circular movements with his scraper, then takes a clean cloth and begins to remove the pasty mixture that has formed. Once only a small amount remains, he takes a clean cloth and removes the residue in straight lines from top to bottom. The window is cleared of its graffiti. José exclaims with satisfaction, 'Well, I thought this one was going to be difficult, but in fact it was easy. You never really know what's going to resist or not.'

By following José, we realize that, in removal interventions, the materials present are not simply understood in terms of their stabilized chemical properties, but as active entities that need to be dealt with. Was acid added to the ink to make it penetrate the glass of the shop window? Does the chemical affect surface texture? José navigates through each removal, feeling his way among active and entangled materials. If the latter are so important and are subject to such careful assessment, it is because understanding their behaviour and what they produce together is an essential part of maintenance activity. Not only do the materials act, but they also react with each other, and it is important that José remains attentive to these reactions. Use of the chemical VG Graff is meant to produce exactly this effect: to make the graffiti ink react so that it liquefies, but not too much, in order to preserve the shop window.

The image of a close physical encounter seems reductive from this point of view, since it can suggest that care involves a face-to-face interaction between one body that is the subject of maintenance and another that is its object. However, as we have clearly seen, many other bodies are involved in this story. José, who was definitely a wonderful guide in our discovery of the art of removing graffiti, insists on this aspect by mentioning the importance of the quality of the rags and the physical properties of the scrapers he uses: 'The choice of rags is important. They mustn't scratch the paint. The recent order placed by [the company officials] hasn't done the job; the new cloths are damaging the surfaces. It's the same with scrapers, by the way. I keep a few really old, tired ones, hardly able to scrape any more, for cases like this.' In the always transformative action of removal, the rhythm and intensity of José's gestures are obviously essential, but the texture of the rags also has its role to play, as does the quantity and degree of concentration of the chemical product he uses and the particular grain of the old scrapers.

José's attitude also gives a glimpse of the emotional dimension of the operation. The latter is anything but effortless. And attention here appears less a matter of diffuse exploration open to the unpredictable than a source of tension, almost of nervousness. In its own way, the scene reveals what maintenance owes to worry. José's intervention could in fact undermine the reason for the existence of the window itself: its transparency. If he is worried, it is because he risks irremediably altering what makes the surface he is dealing with here into a shop window and because he needs, gesture after gesture, material reaction after material reaction, to take this risk into account.

Clearly, this emotional dimension of work is ubiquitous when it comes to taking care of people. It is much less familiar when it comes to taking care of things. However, even in automobile garages, certain interventions are a source of annoyance among mechanics, or even anxiety. Following step by step the actions involved in dismantling the subframe of an old Jaguar, Dant describes the fear that grips Rick, one of the mechanics, as he wonders if he has removed all the parts necessary for the operation while the subframe is suspended and he must act quickly before it comes loose and falls to the ground.[29] The rest of the operation does not lower the tension – far from it. The descent of the subframe is not smooth at all, and the car resists. It gets knocked, and a shower of washers falls to the ground. Rick is petrified to discover halfway through that a piece has stuck and is getting in the way. Fortunately, through a series of coordinated actions, several mechanics guide the subframe down and the manipulation pays off. Once everything is finally on the ground, the relief is great. Fear and anxiety fade into the joy generated by the success of the operation. The frame didn't shatter on the ground; the parts haven't scattered around the workshop. The engine is still there.

A similar tension is found in some of Houston's descriptions of work in Kampala's mobile-phone repair shops.[30] Here's one example: Jason, the experienced owner of one workshop, is approached by a colleague from a neighbouring workshop seeking to restart a phone which remains stubbornly inert despite his interventions. After opening the device and carefully inspecting the components, he eventually suspects that one part is faulty, and delicately de-solders it. He then goes looking for a replacement part in the pile of disassembled phones that litter his work surface. He finds one that he carefully collects and welds it in place of the first. But the phone remains silent. This first failure is followed by two other equally unsuccessful attempts. Jason sends one of his employees to look for another similar phone in the neighbourhood, but the part he managed to unearth doesn't work either. He ends up finding another phone in his pile. He tries once again to perform the transplant. Throughout the sequence, tension increases in the workshop as all those present look at each other in silence; then relief spreads at the sight of the screen lighting up and the confirmation that the object of the care is alive and well again. The installation of the new part is a success: the telephone can remain a telephone.

Obviously not all maintenance situations generate such emotional intensity. Still, the apprehension that we can observe in these scenes, as in many other different configurations, underlines the existential turn that taking care of things can assume. The transformative part of maintenance operates in a continuum of more or less important movements that, step by step, modification after modification, test the very continuity of things and, in doing so, put the people who carry it out in a delicate position. The notion of attachment that we borrowed from Antoine Hennion to describe the dynamics of attention to fragility once again perfectly expresses what drives the worrying side of the sensible maintenance encounters. Taking care of things means being attached to them in a double sense: both holding on to them and being held by them. We have long forgotten it, but this particular tension is present in the very idea of care. This is what the historical dictionary of the French language teaches us about the first uses of the equivalent term *soin* – uses that are not directly associated with the health of people, but evoke a more general sense of concern and preoccupation.[31] 'Caring [*avoir soin*] for something or someone' initially described a propensity for worry or disquiet (*inquiétude*). This disquiet – literally an absence of rest – nourishes maintenance, an art of doing that unfolds in the interstices of material transformations, fostering the sustainability of things day after day.

The dance of maintenance

The same goes for things as for people: care activities are sometimes rough, even brutal,[32] and it would be counterproductive to minimize their violence. Above all, a consideration of these interactions brings about a shift in perspective, rebalancing, reminding us that in maintenance there is no watertight boundary between the world of passive objects on the one hand and the world of attentive and active humans on the other. Taking care of things involves encounters between bodies, in which human beings are not the only ones to act. Certainly, tightening a screw, moving a component or cleaning a part can be sometimes carried out so easily that it seems to be achieved solely by humans, but you have only to observe maintainers throughout their activities to discover that, sometimes just a few minutes after these fluid interventions, a hinge gets stuck, a glue point comes unstuck, or a paint resists solvent. If we could observe in slow motion each maintenance operation, even the most banal, we would in most cases witness this incessant merry-go-round through which the action continues to unfold. Maintenance appears from this point of view as a subtle reciprocal movement, unfolding within the tiny folds of the material exchange of care.

On close inspection, maintenance resembles a dance. And making things last does not so much require mastering their smallest forms of behaviour as learning to live in concert with them, tuning into the appropriate rhythm while paying as much attention to the bodies of each participant as possible. The metaphor is worth dwelling on. It was notably used by Mierle Laderman Ukeles, several of whose performances assumed the appearance of choreography. Whether in big parades, or dances of barges or dustbin lorries, Ukeles got maintainers to let their hair down on several occasions in the streets of New York, Rotterdam, Pittsburgh, Givors in France and Tokamachi in Japan. In these performances there was a desire to display the skills of maintenance workers and to emphasize the specifically cultural role of their activities to the residents and officials who were sometimes brought to join in the ballets themselves. But these works must also be seen as highlighting the practical specificities of the maintenance of large urban infrastructures. By staging these dances, Ukeles shows that the care of things is accomplished in a delicate human and artefactual choreography, where the bodies present create a communal work through the succession of codified gestures and ad hoc improvisations.

The dance metaphor is also useful for refining the way we understand relationships between humans and things. Andrew Pickering has made a very fascinating use of it to underline the part that technical instruments play in scientific practice,[33] without trying to decide whether it is the machine or the researcher that 'genuinely' makes this or that discovery.[34] Once we pay attention to the dynamics of the activity, we realize that it is not one or the other that acts, but *sometimes* one, *sometimes* the other. It's a constant coming and going between humans and things that characterizes the action – a 'dance of agency', in which the different partners alternate between moments of passivity and moments of activity. Sometimes things bend to the will of humans, other times humans obey things.

Tim Ingold has also played with the image of dance to explore the mystery of the relationships between humans and things. He has pushed the reasoning a little further by drawing on several examples, in particular that of a kite.[35] If we follow Pickering's advice, he says, we can completely describe the couple that makes up the kite and the person holding it by paying attention to the alternation of those moments when the hand guides the kite and those when the human at the end of the string reacts to the movements of the piece of fabric or paper, for example by running at full speed to follow its aerial movements and prevent it from being damaged. However, this point of view is relatively unsatisfactory. Why, in fact, stop at this coupling? The example of the kite (but also those of pottery, the lasso or the cello, which Ingold mentions later) clearly shows that some of the elements present are missing. In this case, why, and how, could we remove air from the equation? As long as it was indoors, the kite remained inert. The wind currents play a full part in the flight, at least on an equal footing with the kite and the person holding the string. In Ingold's eyes, it is therefore more accurate to view the dance in question as at least a trio. The demonstration is very convincing and encourages us, as long as we stick to the metaphor, to extend the list of partners playing a role in the dance of maintenance. Sophisticated machines, more or less specialized instruments, screwdrivers, rags, brooms, products of all kinds: we have already encountered them throughout this chapter. We simply have to agree to take them into consideration in order to appreciate, case by case, the importance of their own movements in the collective dance.

It would be premature to enter into the intricacies of Ingold's demonstration at this point. We will return to this later in our own exploration of maintenance situations. However, we can already note a second aspect. By extending Pickering's reflections, Ingold also encourages us to leave aside the idea of a coming and going, a 'lateral' metaphor which still involves too much immobility in his eyes. It is as a *longitudinal* movement that we must appreciate the dance in question, as a deployment of gestures that accompany a music which is never quite the same. This is a very valuable suggestion when it comes to understanding maintenance. We have seen that maintenance needs more than a fleeting face-to-face encounter between humans and things. We have also realized that making things last never means standing still. If maintenance is a dance, it is because it constantly deploys the present to weave the continuity of things by intertwining movements, linking the flow of materials beyond the apparent uniqueness of the instantaneous act and the narrow horizon of mere comings and goings between humans and things. Like a dance, maintenance involves the reiteration of gestures, the multiplication of points of contact, the repetition of patterns. Like a dance, it produces duration from repetition. And like a dance, maintenance finds its raison d'être in the generative part of this repetition, and weaves a continuity which is neither stable nor inert, but rather takes shape in a difference that arises from the repetition itself.[36]

But like any dance, maintenance cannot do without the commitment of the body or the fear of failure, even if this fear is overcome. It is punctuated by friction and feeds on the dancers' anxiety. Criss-crossed transactions, negotiations, reciprocal alterations: this attentive dance, sometimes easy, sometimes tormented, is the incarnate part of the material diplomacy that we evoked in chapter 2 with reference to the Rotor collective's work on wear and tear. Even more than an action, it is the uninterrupted operation through which a *common becoming* is invented and negotiated, on the level of bodies. Maintenance is the long-term dance through which things and people are linked in specific environments whose places, shapes and even compositions evolve through the repetition of steps, always similar, but never completely identical.

5

Time

Nobody knows precisely what interrupted the regular running of the Panthéon clock mechanism in the mid-1960s. Only one thing is certain: from that time onward, until September 2006, the Wagner clock, which is still enthroned today above the famous maquette of the historic building, just opposite the office of the monument's administrator, remained motionless and silent. It is not impossible, in fact, that it was the subject of sabotage, or even of several attempts at sabotage. Indeed, when the original escape wheel was examined, it showed signs of abuse. Had the person responsible for winding it up every week had enough? Had they wanted to free themselves from a commonplace material task that weighed on them so much that they deliberately decided to damage what was supposed to be the very object of their care? Hard to say – especially since, once its original mechanics had been disabled, an electrical system from the Strasbourg brand Ungerer was quickly grafted onto the dial as a replacement. Perhaps some unscrupulous representative of this modern watchmaking company had even sought to ensure that their commercial proposal received all the attention it deserved. These were, it seems, common practices at the time. Sometimes it was necessary to give a helping hand to speed up the march of time, get rid of old things and bring about the era of mass consumption and technical innovation. Adding to the mystery, it appears that this new mechanism, although fitted out with all the irresistible trappings of industrial progress, was in turn the subject of an act of sabotage shortly after its installation. The fate of the clock was then sealed for forty long years, and it was, in this part of the Panthéon, never earlier and never later than 10:49. Nobody, apparently, was bothered. Nor did anyone worry about what had become of the delicate machine designed by Bernard-Henry Wagner, whose movement had accompanied the life of the building for more than a century.

One day in September 2006, Bernard Jeannot, who then held the prestigious position of administrator of the Panthéon, was surprised to see that things had changed. In his office, four people quite unknown to him announced that they had decided to work with some of their friends to restore the clock. Not that they wanted to carry out this thorny operation: they had already carried it out, without anyone in the building, neither its administrative managers, nor the agents responsible for its security, nor the tourists who passed through it all day long, not noticing their presence, and even less the nature of their activities. Thus, the abandoned clock, which had become almost invisible (according to the clandestine restorers, Bernard Jeannot himself had been unaware of its existence before this meeting), had finally received attention. The disused instrument had managed to affect someone, to touch them enough for them to decide, with the help of their companions, to improve its condition and prevent it from disappearing from sight.

The whole business caused a few ructions. However, publicity was not the preferred mode of action of its protagonists: quite the contrary. If we can trust the information they communicated to the press, and then the book that Lazar Kunstmann,[1] their spokesperson, published on the affair,[2] we learn that this was not the first time something similar had been carried out by the four people who met in September 2006 in Bernard Jeannot's office to reveal the secret to him. We also realize that most of the other operations in which they undoubtedly participated will probably never come to light.

Designated by the initials UX (for Urban eXperiment), the collective of which they were a part presents itself as an ill-defined group that came together in the early 1980s around a youthful passion for the exploration of public buildings, particularly underground. In the course of its peregrinations and experimental activities, the group organized itself around specific activities, given more or less cryptic names. In his book, Kunstmann discusses the Mexicaine de Perforation (Mexican Consolidated Drilling Authority), which created artistic events,[3] and House Mouse, a group made up exclusively of women and specialized in undercover operations. One of these activities, designated by the strange nickname Untergunther, is described as a mixture of conservation and restoration actions. We learn from various press reports that its participants took care of a twentieth-century crypt, a century-old bunker, a metro station and even an air raid shelter dating from the First World War.[4] It is difficult to get a precise idea and to separate what were real interventions from the imaginings of journalists and, above all, smoke-and-mirrors operations, an art in which UX members repeatedly claim to be masters. It hardly matters. The episode of the Panthéon clock alone offers a wonderful scenario for discovering many facets of maintenance, including both its conditions of accomplishment and the sometimes dizzying questions that it raises. Let's get back to what we know about it.

While it is undoubtedly fair to depict the members of Untergunther as amateurs, it is important to specify that one of their number, Jean-Baptiste Viot, who had trained in clock restoration, was already at the time of the events a recognized professional, at the head of the after-sales service workshop of a well-established luxury brand. It is to him that the group owed its commitment to restoring the Panthéon clock. As a founding member of UX, Viot had established a special relationship with this monument, presented in Kunstmann's book as the first place that, in 1981, attracted the interest of the very young explorers. Over the years, Viot, passionate about watchmaking (he had started his training in 1983), had made a habit of visiting the Wagner clock, inspecting its mechanism and monitoring its condition. Kunstmann recounts that on the occasion of a show that participants in the Mexicaine de Perforation organized within the monument, the clockmaker returned to see the clock, after several years without a visit. That day, he was struck by how much the pieces had deteriorated. Here's how the scene is reconstructed in the book:

He'd been checking in regularly for over twenty years now – almost every time he passed by, in fact. Only this time, he'd been gone a little longer: almost three years.

'Good God!' he said, speaking to the clock (clockmakers always talk to their mechanisms), 'you're starting to show your age, old chum!'

On the old Wagner clock, the rust had increased dramatically since his last visit. Jean-Baptiste examined it closely, and his diagnosis was even more pessimistic: 'Well,' he sighed, 'you're not going to be around for much longer.'

He had always told himself that, one day, he would do something for this *ancient* mechanism. Seeing it so close to death upset him terribly. Natacha and Lanso, who had joined him, were worried by his expression of dismay. He asked them, 'Maybe we could do something... right?'[5]

They soon came to a decision. In September 2005, the members of the collective began work on restoring the clock. What we know about this project is pretty incredible, to say the least.[6] True artists of camouflage and trespass by both day and night, the clandestine restorers set up a workshop in the circular gallery located at the base of the dome, equipping it with fully retractable furniture and an electrical installation renovated for the occasion. It was in these relatively comfortable premises that the small group worked for a full year, almost daily. The operation first consisted of a massive documentation campaign, intended to facilitate what Kunstmann called the 'autopsy' on the clock. It was necessary to understand the reasons for its inertia. At the end of this time of intensive reading, Jean-Baptiste Viot was convinced that the escape wheel was damaged. At the same time, he formulated the hypothesis that seemed most likely to him: the deterioration was the result of a voluntary act – a sabotage undoubtedly carried out using one of the iron bars intended for the masonry present in the building. The second stage of the project consisted of transporting the mechanism to the workshop. After leaving the clock in a 'clockmaker's bath' (the professionals' trade secret), the team came into physical contact with the materials, in a way very similar to the interventions described in the previous chapter. They rubbed, scratched and patiently polished the parts one by one, and thus gradually freed the elements of the clock from everything that was unwanted: dust, dirt and, above all, rust. Jean-Baptiste Viot made some necessary replacements at the same time. He made a new escape wheel 'with brass as in the old one', explains Kunstmann, as well as two other missing parts. The old replacement electric clock was removed, and the freshly restored mechanism was installed again in its original location, with its glass cabinet. The team also had to change a few pulleys and a series of cables that were no longer in working order.

A year after the start of the project, the experiment had reached a successful conclusion. Untergunther had managed to restore the abandoned clock, whose shiny workings glowed with a new fire in the shadowy heights of the historic monument. But could the operation be considered completed? The answer to the question was far from obvious. In order to understand this, we must accept for a while what the members of this branch of UX have said about their restoration activities. In an interview with art historian Jon Lackman,[7] Kunstmann explains that the collective's interventions do not involve questions of 'functioning' and 'use'. The underground nature, invisible to most people, of the objects and sites they work on is the main reason. Once restored, the things they care for do not have to return to the trials and tribulations of ordinary life, from which they are in some way protected. The Panthéon clock could have met a similar fate. After all, no one had paid any attention to it for the previous forty years, and it was a safe bet that the newfound gleam of its pieces would not attract attention. But a clock is not a crypt or a bunker. It's a delicate piece of machinery whose mechanics are dedicated to movement. It was difficult for Jean-Baptiste Viot and his acolytes to resist the desire to see the hands finally turn, to hear their regular tick-tock. This was not without risk. Starting the clock, and even more winding it regularly, meant bringing it out of an invisibility that it owed in large part... to its silence. With a chime ringing out every quarter of an hour, it is unimaginable that none of the employees of the establishment would notice the change or launch an investigation into this unexpected resurrection. This is how the decision to approach the administrator was taken and the four representatives found themselves in Bernard Jeannot's office explaining to him what they had spent the year on, before taking him to the workshop, then showing him the mechanism in its restored state. The enthusiasm with which he received the news, and the way he lit up with excitement as he listened to the story of the operation, boded well. He seemed like the ideal person to take charge of the clock, make certain that it was wound once a week, and supervise the routine interventions that would ensure that its general condition stopped deteriorating.

But things didn't go quite as this first interview had suggested. In the following weeks, Bernard Jeannot took early retirement and was replaced by Pascal Monnet, his deputy, who was far less pleased to discover that intruders had allowed themselves to occupy the Panthéon for a year. Once in office, he pushed the Centre for National Monuments to file a complaint against the four interlopers who had presented themselves openly to his former superior. However, after several attempts, including a hearing before a judge, no charges were brought and no member of UX was detained. The only drawback, and not the least, was that – as the affair was now public – it was necessary to communicate the details of the operation, which was a first for the collective. The trials and tribulations of the group then became the subject of several articles in the press, including the international papers, and the 'clandestine restorers' became for a time heroes, as it were, of the Parisian underground. Unsurprisingly, it was the hidden nature of the operation that captured the attention of journalists and commentators, fascinated by the UX collective, which they rather simplistically linked, despite the repeated denials of its members, to the world of 'cataphiles'.

What can the details of this quite novel form of restoration teach us, beyond the charms of clandestinity and the romantic idea of this mysterious collective? First and foremost, they confirm many of the aspects highlighted in the previous chapters. The clock benefited from a careful scrutiny, both amateur and expert, that was able to detect the traces of advanced deterioration by getting as close as possible to the material. The clandestine restorers then initiated a series of disassembly and reassembly operations, delving into the tangle of materials present in order to gradually separate the unwanted elements from the original pieces. They finally made the necessary replacements, before delicately putting the whole thing back together. But that's not all. This singular case also puts us on the track of time. And not just because it concerns a clock. It does this by emphasizing an obvious point that we have already expressed many times: to maintain something is to make it last. As long as we don't focus on its implications or its consequences, the expression is rather banal. Nevertheless, what happened, and what did not happen, to the Panthéon clock shows that quite dizzying questions lie hidden in these innocuous terms.

To appreciate the magnitude of the issues at stake, we need to retrace the course of events. Despite the campaign of complaints and the aborted trial, a few months after the meeting between the members of Untergunther and Bernard Jeannot, UX activities within the Panthéon resumed, in particular those of the Mexicaine de Perforation. One Christmas evening, undoubtedly intoxicated by the occasion, several participants asked Jean-Baptiste Viot to wind up the restored clock so that its chime would brighten the festivities. The next day, the new administrator discovered an accurate clock, one that chimed every fifteen minutes. Kunstmann recounts in his book that Pascal Monnet was furious, and immediately hired a craftsman from the Lepaute house to sabotage the clock again. The man was of course incapable of deliberately

THE CARE OF THINGS

damaging the mechanism, and simply removed the main part, the very one that Viot had manufactured as a replacement. The clock stopped again. At first glance, this new episode marked a resounding failure. First by exposing themselves to Jeannot, and then by restarting the clock for a few days, the members of Untergunther had shown themselves to be both too naïve and too arrogant, and the very object of their attention had ended up paying the price for their imprudence. But to conclude that their efforts were rendered pointless by this new act of sabotage would be to miss the fundamental principles defended by the collective. Far from the clichés of the general press, Kunstmann had managed several times to present the action of Untergunther[8] by linking it to a highly structured vision of heritage conservation.[9] As we have already mentioned, he first stated that the operation's aim had not been to get the hands of the clock turning again. The purpose had mainly been to prevent it from disappearing: 'Untergunther's main concern was not so much that the clock would work again, but that it should be saved from irreversible deterioration.'[10] The maintenance carried out by the collective was therefore a matter of life or death. By discovering the state of the mechanism, Jean-Baptiste Viot understood that the moment of disappearance was imminent and convinced his friends that urgent action was necessary. The restoration of the Panthéon clock was a last-chance operation: a matter of now or never.[11] Thus, only one thing mattered from the collective's point of view: that the object of their care had not disintegrated. The clock remained in place, *present*. It had saved time. It had lasted, and it still lasts. That said, by intervening, the members of Untergunther had not only sought to stop the process of deterioration. They had striven to erase the very marks of ageing from the clock (dust, rust spots, etc.), they had replaced certain elements, rebuilt a part of it... What was at stake here was a very particular way of making something last. As if the restorers had sought to go back in time, to return to an earlier stage of the clock's life.

While the clandestine restorers are very striking figures, their position should not be taken for granted, nor seen as the only correct way to extend the life of the clock in question. Let's return to Pascal Monnet, the new administrator who was so hostile to Untergunther, and put ourselves in his place for a moment. What exactly was he doing when he ordered a craftsman to stop the clock's mechanics from working again?

What did this second act of sabotage accomplish, if not also a 'step back in time', an *alternative* step back in time? If we take his gesture seriously, as we took that of the clandestine restorers, we can conclude that, by once again depriving the clock of its escape wheel, Pascal Monnet had established himself as a restorer competing with Untergunther. A restorer who simply wanted to go back a little less far into the past and who, rather than starting from the nineteenth-century clock, wanted to reinstall the unobtrusive clock, without fuss, that had been enthroned in the building from the mid-1960s until 2006. With his own intervention, Pascal Monnet brought in another way of making the clock last and, with it, the entire monument for which he was now responsible. By taking the argument to its logical conclusion, and keeping in mind that part of the mechanism was removed, we can even say that Monnet made another clock last: 'the stopped clock of the Panthéon',[12] the very same one that had been there on the premises for more than forty years, an integral element of the monument, immobile and silent. By interpreting Pascal Monnet's position not as a new act of sabotage, or even as a rejection of restoration (something for which the members of Untergunther criticized him), we realize that making something last is also quite simply 'making it exist'. And if there are several ways to make things last, it is likely that in maintenance we will find several ways to make them exist. As each of them is tied to distinct times, the restored versions of the Panthéon clock enacted by Untergunther and Pascal Monnet encourage us to pay attention to the ontological dimension of the care of things.

If the speculative exercise of comparing and contrasting these positions is worth the detour, it is also because by highlighting the differences, it stops us being tempted to make duration itself a univocal operator linking past and present in a transparent manner. This is what Antoine Hennion tells us in his discussion of artistic restoration.[13] Once we examine the gestures and practical concerns of the restorers, and even more so if we focus on their evolution and their quarrels, we cannot fail to be struck by the generative character of the activity of restoration. Far from developing operations that would mechanically allow 'going back in time' in order to rediscover a past crystallized in objective properties that can simply be made to happen again, restoration is a 'workshop of history' within which the relationship between present and past is literally manufactured: 'Few concrete activities are as prone as restoration to constantly "make" history, in the most material sense that the expression here assumes.'[14] Time is thus the material worked on by restoration, which amounts to producing the past (*a* past, we should say), by ensuring it particular modalities of presence. What goes for restoration and conservation goes for any form of maintenance. By making things last, we always manufacture history.

This is what is striking in the contested episode of the restoration of the Panthéon clock. Of course, these are two 'visions' that clash, two rather abstract historical horizons; but above all, it is ways of *doing*, forms of organizing action, that have been competing with one another. To make something last is to tackle specific tasks in order to work on the time of that thing.

Thus, according to Untergunther, if there was need to act it was because the time of the abandoned clock was a destructive time, a time in which each element of the mechanism was gradually being devoured by rust. The time around which the new administrator's action was organized, on the other hand, was a stopped time, a return of the clock to its 'normal' state of the 1960s, one that the clandestine restorers had disrupted. If these two ways of considering the clock's time are profoundly incompatible, it is because their practical implications differ drastically. Once the new escape wheel had been removed, Pascal Monnet had accomplished his work and no longer needed to worry about what would happen next. Time could once again take its course. However, if the members of Untergunther snatched the clock from the imminent time of its disappearance, they were not merely leaving it to its fate. As Kunstmann wrote in 2009 in the epilogue to the first edition of his book: 'The rest of the clock has, right from the start, resumed its process of oxidation.'[15] The 'step back in time' and the relationship thereby established with the past are not thought of as definitive, and there is no doubt for the clandestine restorers that further operations will be necessary. Actually, other members of UX were quick to steal the new escape wheel confiscated by Pascal Monnet in order to put it 'in a safe place', so that it could one day be used again. So much the better for them: in 2018, the French Centre for National Monuments, through the offices of Christophe Niedziocha, the curator, and Gaëtan Bruel, Pascal Monnet's successor, finally repositioned itself on the timeline that Untergunther had striven to update by launching a vast restoration project for the entire clock (and no longer just its mechanism). The escape wheel returned to its place, since this part of the worksite was entrusted to one of those people who knew where it was: Jean-Baptiste Viot himself.

The different forms of inscription in time that clash in the case of the Panthéon clock suggest the political nature of making things last. In the care given to objects, small or large, certain ways of 'making history' and the practical conditions for keeping things going are defended together. These conditions shape the pattern of attention to fragility but also provide the benchmarks necessary for the more or less transformative interventions of the maintainers. How can we account for these contrasting times? How can we understand the implications of their differences? It would be tempting to answer these questions by launching into a systematic programme, identifying a few rigid temporal 'frames' that we would organize according to well-chosen examples based on principles, rules and tools. We could then establish a grammar of maintenance that would redistribute its temporalities in a well-ordered table and summarize the diversity of practices in a set of overarching 'conceptions'. But that would mean completely failing to recognize the richness of the situations. The dialogue between different cases that we embarked on in our earlier chapters showed how varied are the issues and how subtle the nuances. Above all, it helped us understand that maintenance was a matter of inquiries, in the strong sense of the term: a matter of doubts, explorations, concerns. This is where we need to start from, in order to continue our exploration of the temporalities at work. The case of the restoration of the Panthéon clock showed us that the temporal perspectives deployed in the ways of making things last do not represent competing strategies aimed at providing distinct responses to a common general problem, but specific ways of seeing time as a problem – a problem that is simultaneously political, moral, sometimes conceptual, and always practical, even commonplace.

All things considered, this perspective is in line with the approach that François Hartog developed when he began exploring 'regimes of historicity' as a historian. In his book, he looks at the practical experience of time by documenting its evolution and transformations.[16] In doing so, he breaks with the analytical reflex which makes time a principle of external explanation, an objective plane on which are projected facts that, by this very gesture, become 'historical'. By identifying regimes of historicity, Hartog makes time itself an object of history. This is an ambitious and fruitful manoeuvre and leads to writing a history of history, or rather a history of histories, where regimes succeed one another within which specific relationships to the past, future and present are articulated. Because it encourages us to become attentive to the way in which time poses a problem for people who practise maintenance, we propose to adopt this as an approach, bringing it down to ground level, to the operations carried out to make things last. And precisely, as soon as we try to understand how, and in what way, time poses a problem in maintenance, we realize what a delicate operation it is to make things last. Firstly, maintenance never boils down to a generative process that simply brings about the time of the thing. Maintenance also involves a form of destruction. If it enacts a duration that it actualizes under certain conditions, it is because it 'fights' against other possible, undesirable times. The forms of maintenance therefore connect two problems: that of a time that must occur and that of a time that must be resisted. But as we mentioned in our discussion of the competition between the two versions of the Panthéon clock, an approach based on the work of time also highlights the ontological dimension of maintenance. Taking care of something, ensuring that it lasts, always comes down to selecting, from the multitude of traits that define it, those that we value and whose existence we wish to prolong – those that, precisely, will ensure that it is indeed the 'same' thing that lasts. Conversely, in this movement, many aspects of the things involved will be neglected, or simply ignored. This ontological operation is generally buried in the ordinary routine of maintenance. It seems an obvious part of the mode of existence of many things. It is only by comparing concrete situations that we can measure its importance and grasp the depth of what the expression 'to make things last' means.

Throughout our own investigations and through our meetings with colleagues whose work has gradually nourished and consolidated the field of maintenance studies, we have learned to be sensitive to the diversity of ways of making things last, and even more to the variety of ways of problematizing time. We have come to identify four forms of maintenance, within which modalities of confrontation with time are connected with ontological gestures: prolongation, permanence, slowing down and stubbornness. Because they are always based on concrete situations, these four terms are not intended to form a system, and we will not seek to set up a systematic comparison of their characteristics. It is the contrasts generated by their serialization that interest us. Given the dynamism of the field of research and the wealth of studies now being published at a sustained pace, we have no doubt that other forms can be described.

Prolongation

Let's start by looking at the most mundane form of maintenance, the one that is practised every day and at first glance touches only superficially on the torments of our relationship with time. In fact, it is rare for it to be explicitly identified by the term 'maintenance'. It is rather called 'upkeep', or not labelled at all. One practises it without really thinking about it, like many of the situations described in our introduction. Let's take a car, any kind of car. How does one try to make it last? What does its maintenance comprise? These questions can be answered simply by drawing up a list of operations that are considered successful from the moment they become routine actions. Checking and adjusting tyre pressure before each long journey. Replacing the tyres when they are worn down. Regularly changing the oil, the oil filter, the air filter and more rarely the cam belt; carrying out a systematic service from time to time to ensure that the main parts are in good condition and that everything is working as well as it can. Carrying out necessary replacement work, and repairing damaged elements after a collision. There's nothing metaphysical in these different tasks. While some require advanced technical knowledge and specialized tools, they remain quite commonplace. In terms of temporality, it is rare that they lead those who carry them out to make any subtle reflections. It's a very minimal form of 'making things last' that's involved here. We are not seeking to rediscover a more or less distant past, nor do we project an inordinate ambition towards a future that we seek to bring into being. In this form of maintenance, time only poses a problem in fairly trivial terms, without any conceptual or instrumental sophistication being necessary. By maintaining our car, we ensure that it lasts 'a little longer', that it simply continues to be a car.

In simple terms, what is at stake here can be considered to be a matter of prolongation. Prolonging the existence of things is a daily, almost insignificant experience. It's what we do when we polish our shoes, when we have the boiler serviced, when we clean the dishwasher filter, when we have the battery of our smartphone replaced... It's everything that can be brought together under the imprecise label of routine maintenance. It's also the quintessence of the art of maintenance that Mierle Laderman Ukeles celebrates, in all its monotony and the lack of ambition of its temporal horizon. Domestic life has long been the epicentre of this maintenance, one of the paradigmatic gestures of which is mending. Much less widespread today, this practice is far from having disappeared, particularly in homes where there are young children – those past masters in the ability to wear out clothes. It is an 'art of making' that is often transmitted from generation to generation, although never celebrated. No one pays attention to it and few people view it as a militant action.[17]

But it is precisely because it seems so bland and so unpresumptuous that we must explore the diffuse time of prolongation; starting out from this, we will discover how time, in other situations, poses very different problems. What does prolongation teach us? It first suggests the importance of a well-known distinction, one that is not always easy to make but seems to play a crucial role here: prolongation, a matter of light-touch maintenance, is much more concerned with *use* than with *form*. We can of course deal with both requirements, trying to reconcile an optimal functioning with the demands of authenticity or aesthetics. But it is not to this aspect of objects that the gesture of prolongation is intimately attached.

What drives it above all is the need to avoid breakage or breakdown, things that would interrupt the thing's ordinary life cycle. The minimum objective is that these interruptions should be temporary so that we can quickly use the object again. Such an attachment to use implies significant consequences for the shape of the thing whose life we seek to prolong. It goes hand in hand with a relative disinterest in formal integrity. Nothing, for example, stops us replacing the door of our Renault Espace, dented by a clumsy fellow car park user, with a door recovered from the scrapyard, even if it's not exactly the same colour as the rest of the car's body. The passenger compartment will still be able to close, the car will be fine to drive and the outcome of the forthcoming MOT won't be any different. In other words, this form of minimal problematization of time goes hand in hand with a sort of 'ontological relaxation' which accepts a significant degree of transformation. This is particularly what Nicky Gregson and her colleagues demonstrate after a long-term investigation that they carried out in the north-east of England.[18] On examining the usual consumption practices of the residents of a small town, they were astonished by the variety of types of maintenance that the latter carried out. These included the temporary repair of a TV set by quickly disguising a crack, and the successive repatching of the leather of a worn sofa. They show that, for some people, prolonging the life of an object mainly consists of ensuring that it always fulfils its main function, even if it means that its physical integrity is visibly altered.[19]

The wife of one of the authors of this book found a way – a good example of this ontological relaxation – to prolong the life of their youngest son's jeans. As soon as the skin of the boy's knees was visible through the threadbare, or even torn, fabric, she began a daring but effective destructuring operation. She unstitched the back pockets of the jeans and sewed them onto the rips, providing knees a little too prone to rubbing against the asphalt of the schoolyard with long hours of extra protection, without having to buy new jeans every month. No one was offended by this sleight of hand – one that in other situations, on other types of clothing, might be considered a sacrilege, an irremediable act of denaturalization. All that mattered here was that the jeans could be worn again without the child hurting his knees, and probably without his parents being looked at askance at the school gate.[20]

This disinterest in form is not mandatory, of course. It is not a condition of the ordinary maintenance which only aims to prolong the existence of things for an indeterminate time. However, the logic of prolongation bears this possibility within it. This is particularly visible in countries where the need to prolong the day-to-day functioning of a large number of things is vital, those countries which make up the 'poor world', which the historians, sociologists and economists of technology have ignored for decades.[21] We have already discovered the extent of the disassembly operations and modifications that repairers in Kampala workshops carry out on certain phones so that these can keep working for a few more months.[22] It is these few months that mainly matter here, and certainly not the respect for any material, aesthetic – or even functional – integrity, since in the name of this prolongation, some of

THE CARE OF THINGS

the initial technical possibilities can be abandoned in favour of others judged more essential.

By the relatively unproblematized relationship to time that it fosters, as well as its ontological relaxation, this form of maintenance, the most ordinary and most frequently practised in the world, provides a mine of counter-examples capable of shaking to pieces the great frescoes which usually depict the role of technology in the history of humanity. Prolongation calls into question the evolutionary view that sees the West as the sole focus of the history of technology.[23] This analytical template, which is still very largely dominant nowadays, is absolutely blind to the multitude of technical objects whose life is continually prolonged while their forms, their functions and their uses are tirelessly and discreetly transformed, to the point of becoming 'creolized' by dint of adjustments, borrowings and adaptations.[24] The life of humans in the North and the South, in the East and the West, is nevertheless populated by these 'prolonged' objects, which form an essential part of the socio-technical fabric of the world. And, as soon as we take it into consideration, their very prolongation, the ill-defined temporality that unfolds modestly, with a light touch, provides the material for another history and another geography of technology.

It must also be recognized that, despite its trivial nature, the quest for the prolongation of certain things can assume overtly political forms. This is demonstrated by the movements fighting 'planned obsolescence' and militating for the 'right to repair' that have grown in recent years. We will have to return to the history of these militant actions, which deserve a more detailed discussion.[25] Let us simply note for now that they are characterized by the refusal to see large companies impose an arbitrary limit on the lifespan of the consumer goods that they place on the market. The practices denounced are in particular those which purely and simply prevent users and independent repairers from carrying out the most ordinary operations of prolongation.

In certain economic configurations, the gesture of prolongation, as banal as it may seem, is therefore a political gesture that places the undecidability of the time of the thing at the centre of practices. Making something last 'a little longer' amounts to clearing the temporal horizon of modernity once more and endeavouring to keep it open, even when circumstances might suggest that it is definitively blocked. This shows, in passing, the extent to which the idea of 'breakdown' so often deployed by the social sciences as a way into the 'black box' of technology deserves to be modified and reconsidered, as we explained in the first chapter. The political operation of prolongation is an affirmation in practice of the relative, and therefore negotiable, nature of the lifespan of objects, and of breakdown itself, even if this was planned.

That said, it should also be noted that while this form of maintenance can translate into vehement protests and be organized into highly structured forms of struggle, it weaves into the care given to things a temporality characterized by its absence of pretension. The operation, the form of openness that it promotes, is not trivial – far from it. However, it remains modest, entirely circumscribed in this 'just a little bit more' of prolongation, which says nothing about what happens next. To maintain, here, is to escape *for a moment* from death, from breakdown, from uselessness, from disappearance or being definitively shunted off towards the world of waste. But about this moment, we don't know much; we do not claim to know its boundaries. We'll see.

If it is useful to try to identify the particular relationship to time that this first way of taking care of things, prolongation, fosters and to underline its political implications; this is because other forms of maintenance stand out from it significantly. A good way of grasping the differences hidden behind the blinding obviousness of the expression 'to make things last' is to start from the question of pretension. Let's imagine drawing a continuum on which to place practices of maintenance according to their ambition, a sort of scale of modesty that would distribute the temporalities they create from the most humble to the most presumptuous. This interpretive template is a bit simplistic, of course, but it is useful for generating a first comparison. Thus, we would find on this gradient, opposite the very reasonable idea of prolongation and its pragmatics of 'a little more', maintenance activities that focus on making certain things last 'forever'. These operations, generally structured in explicit and official programmes, are part of a problematization of time no longer expressed only in terms of lengthening, but in those of permanence, even eternity.

Permanence

To understand how this temporality is enacted and to highlight the problems it poses, we propose to look in detail at two singular cases which in principle are completely opposed: the preservation of the body of a politician who is among the greatest figures of the twentieth century and the maintenance of Paris metro signage, several aspects of which we have already depicted. In addition to the obvious pleasure of provocation, the successive exploration of these two cases, at first glance fundamentally antagonistic, will help us identify the main features of a maintenance that is organized in search of permanence. We will see that beyond the specificities of social, material and historical configurations, making something last 'forever' goes hand in hand with a certain configuration of maintenance activities and is based on an ontology of the thing maintained (its authenticity, its integrity) that is far from obvious.

Let's first delve into the most singular and surprising story that we discovered in our investigations into maintenance: the preservation of Lenin's body, still on display today on Red Square in Moscow in a mausoleum which adjoins the Kremlin. Alexei Yurchak, who carefully studied the scientific and political adventure of the conservation of this body, will serve as our guide.[26] We could stick to a symbolic analysis of Lenin's remains, the conservation of which would then be considered solely through the prism of their cultural value (their function as the 'representation' of a certain ideology). But the great richness of Yurchak's work lies precisely in his rejection of such reductionism and his analysis of the concrete conditions of preservation, which allows him to grasp the political implications and the most trivial material aspects at the same time.

Let's examine the sequences that have marked the life – so to speak – of Lenin's remains. The preservation of the body had initially taken a traditional turn. Lenin was the subject of a completely classic embalming, before being exhibited for six days at the House of Trade Unions, where a gigantic crowd came from all over the country to pay homage to him. At the end of this conventional stage in proceedings, the body was placed in a glass coffin intended for burial. The crowds kept streaming in, so those responsible decided to display this coffin for a few more days in a mausoleum on Red Square before the actual burial, scheduled a few weeks later. The days went by. The winter was harsh and the temperatures exceptionally low. Lenin seemed to be holding on pretty well. The situation continued, in an unprecedented and completely unforeseen way. A medical commission was formed, which examined the body regularly and ensured that it did not show signs of decomposition. He remained like this, in a sort of intermediate state, until 26 March 1924, two months after his death, when the first signs of physical deterioration were noticed: the colour of the skin changed in places, and certain tissues softened.

This interstitial period of time, which no one had scheduled and which was in some way imposed on everyone, created a window for discussions among the Soviet authorities. Once the usual monitoring period had passed, the procedure to follow was in fact no longer obvious, and each decision was the subject of debate, sometimes heated, within the Commission for the Organization of Lenin's Funeral. One of the delicate questions that arose was naturally the question of the status of this public exhibition and its prolongation. Some people immediately underlined the theoretical and political impossibility of transforming Lenin into a relic, as religions do with their saints. This would be to directly contradict the materialist principles at the heart of the theoretical apparatus of communism. On the other hand, it was difficult for the members of the commission to ignore that the population was still continuing to arrive. Exhibiting the body a little longer so that everyone could say goodbye to Lenin was eventually viewed as an acceptable possibility. But how was this to be done? The condition required by this option lay, of course, in maintaining the body as it was. There was no question of exposing a decomposing corpse.

Although it only temporarily postponed the moment of burial, the decision significantly modified the trajectory of Lenin's body by disrupting the environment in which it had been located until then. Around the remains of the former Party leader, numerous scientists from various disciplines appeared, mandated to tackle the question of preservation which, until then, had not really arisen. Indeed, the body, the mausoleum and the Russian winter climate had proved to combine in a way almost sufficient for conservation. Many proposals emerged on this occasion, but none made it possible to prevent the appearance of the body from changing rapidly. At the slightest sign of decomposition, it was understood that the body had to be buried. However, over the course of the debates, the imaginable temporal horizon for its conservation gradually widened. Although no technical solution was available to the authorities at this stage for it to be implemented, the idea of a form of longer-term maintenance of the body nevertheless took hold, little by little. Eventually, a second commission was created, formed without the first being revoked: the Commission for the Preservation of Lenin's Body. It was within this new commission that at the end of March 1924, when the first signs of deterioration were appearing, it was decided that a new method would be tested. This one, designed by a doctor and a biochemist, relied on recurring baths and regular injections of chemicals. Four months later, as the frozen body had again held up well, Vorobiev and Zbarsky, the two scientists, announced that their tests had been conclusive: the method could ensure a long-term presence for Lenin's body, even a very long-term presence: potentially eternal.

Thus, from January to July 1924, it was not only the list of people involved in the preservation of Lenin's body that changed, but the timeline in which he was located that was gradually reinvented. The conventional parenthesis of a few days of mourning, then the weeks gained with the help of the glass coffin and the cold, was replaced by a much broader horizon, a permanence clearly evidenced by the new name given to what was now one single commission: the Commission for the Immortalization of Lenin's Memory. Almost a century later, Lenin's body, which has become a carefully maintained object of memory, continues to be displayed to the public.

Let's leave Lenin where he is for the moment and change the scene radically. Let's return to the corridors of the Paris metro and take a closer look at the signs that guide travellers in their underground wanderings. We talked about how our own views had been transformed by the experience of shadowing the maintainers of the RATP signage and how the material fragility of those signboards, that we had imagined to be indestructible, had ended up staring us in the face. We also briefly mentioned the vast standardization operation that led to the installation of this Parisian signage in the 1990s. This operation saw each of its graphic, material and language components standardized down to the smallest detail and designed as an element of a system of interdependent signs. Beyond the striking discrepancy between the fixity intended by the designers and the deterioration that maintainers face on a day-to-day basis, we realized that it was necessary to understand the two aspects symmetrically, as two inseparable dimensions of the same world. Far from contradicting the designers' project, the maintainers' attention to fragility was in fact at the service of the very existence of the apparatus the former had created. It is the maintenance of the signage that ensures its stability in practice, and *realizes* the programme for the graphic ordering of transport spaces – a programme that the material properties of the signboards could not carry out on their own.

In terms of temporality, the link between ambitious design and attentive maintenance created the conditions for a permanent signage in the Paris metro. However, unlike Lenin's post-mortem tribulations, the possibility of permanence here is not circumstantial. From the outset, it lies at the heart of the system whose all-round standardization goes hand in hand with a model of omnipresence. Indeed, it is one of the elements that characterize the ambitions of the signage designed in the 1990s: the wayfinding system is organized around a maximalist principle that led to an increase in the number of signboards in the stations and bound the operation of the entire system to their continued presence. This omnipresence is a political gesture on the part of the transport company. It is one of the expressions of a certain idea of mobility, envisaged as a universal service provided to all metro users. To the 'engineer's' model of transport, which consists of ensuring the movement of a person from point A to point B, this service model adds a preoccupation with comfort. In the mid-1990s, transporting travellers was no longer enough; it was now necessary to take their experience into account and do every-thing to improve their conditions.[27] One way to operationalize this new requirement was to provide RATP users with signs everywhere and all the time. The numerous internal documents which accompanied the instal-lation of the Parisian signage presented it as equipment for facilitating travel, intended in particular to 'reassure' travellers who had every reason to be anxious in an underground environment. The permanent presence of this feature, ensured by daily maintenance, makes it an instrument on which everyone can rely, including the most vulnerable traveller.

As the reader will agree, bringing together the history of the progressive 'eternalization' of Lenin's body and the establishment of omnipresent signage in Parisian transport spaces is a somewhat delicate, indeed frankly far-fetched, operation. One can draw up a long list of the things that separate them: the differences include the things that need to be taken care of, the circumstances in which a programme for permanence ends up being imposed, the materials concerned, the maintenance methods developed and the political projects that they embody. However, when examined from the point of view of maintenance and understood as ways of creating a particular temporality, the few points they share still have a lot to teach us.

One first element deserves to be underlined. Even if we are dealing with explicit programmes and detailed strategies, in both cases the permanence of things is not self-evident. Their long-term duration is not obvious – it is not something that could constitute the starting point of operations or their motive. It is only imposed at the end of a process which seems to make the various people concerned aware that this kind of relationship with time is possible, even necessary. This is obvious in the case of Lenin's body: Yurchak notes that it was unimaginable in the first weeks following his death to envisage long-term conservation, which was even considered at first glance as incompatible with communist ideology. It was only through more or less fortuitous episodes that Lenin's body could be understood as a memorial object intended to perpetuate a certain version of Leninism, an object that a complex and completely new scientific method would make it possible to preserve in all its vivacity. And it was through the practice (and in this case the science) of maintenance that permanence became a possible horizon. As far as signage is concerned, awareness emerged from an almost symmetrically opposite process. Here, on the contrary, it was the initial absence of maintenance that made it imperative to make the system permanent. For while the original programme had been organized on the basis of consistent signage components and their omnipresence in transport spaces, it had not planned anything for the system to continue. Permanence was in fact a central postulate of the conceptual architecture that guided the creation of the new signage, but it remained implicit. Above all, no one had seen it as a problem, a property that needed to be created. And this remained true until the first metro lines were equipped with the signboards in question... and these signboards showed some of the signs of deterioration that we have mentioned: discoloration, rust, breakage, or even complete disappearance. It was only when these first failures occurred and came to the notice of the highest echelons of the company that the time of the signboards became problematic and that this problematization was translated into terms of maintenance. It was therefore necessary to wait until 2000, three years after the first wave of installation of the elements of the new signage, for a maintenance department to be created, specifically dedicated to the maintenance of this 'mobility assistance equipment'. It was as if, until then, no one had realized that its perpetual presence, implicitly considered crucial to its effectiveness, required it to be ensured on a daily basis.

While it can be remarkable in its wide scope, the relationship with time that is established in both situations is therefore the result of inquiries, doubts and sometimes trial and error. It is not set up as a maintenance 'model' whose modalities are identified in advance; rather, it is experienced simultaneously as a political and material problem.

Over and above their differences, the two cases also reveal a second important aspect of this manufacturing of permanence: the incessant work that accompanies it. The perpetual fixation of things occurs through a series of constantly repeated operations, accomplished by maintainers whose uninterrupted activity contrasts with the immobility that they strive to generate. Thus, at the time of our investigation, around fifteen people worked full-time in the signage maintenance department of the Paris metro. Every morning, at 7 a.m., the manager distributed to two pairs the list of interventions to be carried out during the day. Their trucks then criss-crossed the city from one station to another, each team sometimes replacing a damaged or missing sign, sometimes measuring an element to be changed, assessing the state of deterioration of a sign that had been reported or even adjusting the fixtures of another. Before that, the signboards had already received a visit from the agents responsible for the stations, who had surveyed the spaces before their opening to the public, in order to monitor possible transformations, then had transmitted requests for intervention to the departments concerned – like Nadine, whom we saw at work in chapter 3. As well as these two daily rounds, there were the people who worked in the maintenance workshop, designing and manufacturing the temporary PVC boards and preparing the orders for their 'permanent' versions in enamelled sheet metal, produced by service companies. The regular and continual movements of this army of workers (which may also seem very modest compared to the thousands of panels which punctuate the territory of the Paris metro) constitute the cogs of a sort of mechanism of inertia. Each intervention operates within a vast machine for the production of permanence, deployed all around the signboards.

THE CARE OF THINGS

It is easy to imagine that an apparatus such as signage benefits from this type of machinery – even if, as a metro user, it's not what you first think of. Its existence around Lenin's body, on the other hand, may seem much stranger. However, it is indeed by becoming the object of a complex machinery that Lenin's body was able to claim a stake in eternity. And the least we can say is that remaining an immutable body is no easy matter. Clearly, far from offering a miracle treatment that would stabilize the body once and for all, most of the scientific methods that Vorobiev and Zbarsky developed were organized into a series of varied operations, arranged in a strict protocol that has been modified and extended over the years. Prolonged immersions in chemical solutions, injections of substances of all kinds, regular 'touching-ups' of surfaces, the remodelling of shapes... here too, the flip side of fixity turns out to be a matter of endless agitation. These interventions are standardized and distributed in procedures whose rhythm varies. The longest treatment, which can last up to sixty days, has to be repeated every eighteen months. Others are monthly, others weekly, and some treatments must be carried out daily. This list shows that the permanence produced in this configuration is in some way intermittent. It is punctuated by parentheses, some of which require that the public exhibition of Lenin's body be suspended in the name of its perpetual renewal. What is perhaps even more striking in Yurchak's description of this machinery is the extent of the human resources that were devoted to it. We have seen that the first moments of hesitation after the initial phase of exhibition of the body had been an opportunity to undertake urgent scientific explorations. These had led to the stabilization of a first methodology considered convincing enough for the process of 'eternalization' to begin. But things didn't stop there. The preservation of the body was the focus of intense and continuous scientific activity, which brought together a much larger army than that devoted to the sustainability of Parisian signage. Thus, in 1939, a first research laboratory was created, which later became a full-fledged research centre. Although, from the end of the 1980s, the number of people associated with it decreased significantly, the centre was able to assemble up to 200 researchers at its peak, during the 1960s and 1970s. Fabricating the permanence of something like the body of a human being so that it could become the incarnation of the memory of a political ideology involved a completely new range of activities. Yurchak also shows that the latter proved crucial for certain scientific advances, despite the complete secrecy in which it was shrouded.[28] Today, it still represents a real challenge, halfway between science and art.

In order to better understand this form of maintenance, one that poses the problem of duration in terms of permanence, we can therefore call on the image of a thing made eternal by a multitude of necessarily ceaseless small interventions. Here, the two phases of maintenance are closely linked: the sustained and regular phase of the maintainers, and the peaceful and infinite phase of the thing maintained. The agitation of the former is entirely devoted to the immobility of the latter. Maintainers work constantly so that nothing will happen, no transformation will be perceptible and the thing they take care of can last every day, the same forever. We saw in the previous chapter that the metaphor of dance was a valuable way of grasping the dynamics of material encounters in maintenance, through which the bodies of people and things alternate between pre-established repetitive patterns and improvisations. By picking up the thread of this metaphor and rounding out the ideas that Mierle Laderman Ukeles has set in motion in her dance performances, we can grasp the choreographic particularities of a maintenance that looks towards eternity. Rather than a ballroom dance or a ballet, we are dealing here with what could be compared to a scene from a musical comedy in which the thing plays the role of the central character and the maintainers flutter around, their movements exclusively dedicated to putting the star in the spotlight – except that, in this case, the action onstage produces a very particular effect on the audience, similar to that sought by certain street artists who disguise themselves as statues for the tourists to admire: the effect is an impression of absolute fixity. Instead of enhancing the grace of the star's movements, those of their support performers are so arranged that we see simply a smooth, unchanged image, and the star's slight quivers, the inevitable microvariations of their body, remain imperceptible to the majority of people. Think of the Eiffel Tower, a huge, impassive lady who has become the symbol of Paris in her own right. Her postcard-like confidence, her apparent immutability, are the result of a permanent ballet, ensured in particular by the fluttering of painters who continue over the years to maintain the very particular tint of the iron of her dress, following as it were a musical score, floor

THE CARE OF THINGS

by floor, year by year, until the layer of paint gets too heavy and has to be removed, before starting over and over again.

There is one more point in common between the otherwise incongruous phenomena of the maintenance of the Paris metro signage and the preservation of Lenin's body. It can be stated in the form of a question: what exactly is being maintained? An impatient reader might be tempted to respond in annoyance, 'Come on now, you've been harping on all this since the beginning: signage, and Lenin's body'. True enough. But sticking to these two terms, ultimately very abstract, would not help us understand the ontological gesture of a maintenance that aims for permanence. What matters in the Paris signage and in Lenin's body in the eyes of those who make them last every day, forever? What definition do they rely on so as to actually carry out maintenance? We have seen that in cases where it is essentially a question of prolonging the existence of a thing without making long-term plans, a large number of modifications are possible. Replacements, additions, alterations: the features of the thing are sufficiently flexible in this form of maintenance for all these interventions to be carried out without any big worries. Even if for different reasons, it is obvious that such flexibility does not apply in the two cases that we have just explored. Permanence cannot be obtained at just any price. Since it is a question of producing a certain fixity, such an open-ended approach is ruled out. Certain properties have to be met, which ensure that the thing remains 'itself'.

However, and this is perhaps one of the most surprising lessons of this form of maintenance, we have sensed that the requirement for fixity is combined with a modular definition of the thing being taken care of. This is obvious in the case of metro signage. There's nothing dramatic about replacing a signboard: on the contrary, it is through this gesture that the permanence of the entire wayfinding system is ensured. As long as you do it in compliance with standards, that is to say by installing a signboard identical in every way to the one it replaces. If each station manager came up with the idea of replacing a damaged sign with a sheet of paper on which an arrow and the letters forming the name of the direction were drawn by hand, or an enamel panel displaying a typography other than the regulation Parisine,[29] the system would be considered by those in charge of it as faulty, and its maintenance as defective. Signage here is therefore conceived of as a generic thing: a set of modules that are carefully made to coexist uninterruptedly, but that are regularly replaced, for the sake of the permanence of the whole, itself dedicated to a service that the company provides for its users.

One might believe that Lenin's body is subject to a very different regime, its maintainers responding to a much more drastic requirement for its integrity, if only because they are dealing with a biological entity. However, it is also on a certain modularity that its eternity is based – a modularity that contradicts the physiological definition of the integrity of the body, and continues to intrigue people after decades of conservation. As Yurchak explains, in the 2000s a deputy from the Duma protested against the expense of preserving Lenin's body. In order to defend his position and win the argument, he sought to demonstrate that the very quest for eternity was based on an artifice. He claimed that, after all that time, it was probable that there was almost nothing left of the 'real' Lenin in the body resting in the Red Square mausoleum, 'barely ten percent'.[30] Taking the question seriously and deciding to play the game of percentages, the journalists of a weekly newspaper began calculating the odds, based on the various documents they were able to consult. Without confirming the deputy's statements, the paper published a result that was enough to leave one wondering: it could be concluded that the exhibited remains were no longer composed of more than 'just twenty-three percent' of the biological material that had once made up Lenin's living body. Nothing surprising in that, really. As we have already mentioned, many internal organs had been removed, while liquid and solid substances had been regularly injected into the body for years to ensure its preservation. These oft-repeated changes and replacements gradually established themselves as the necessary conditions for the fixity of the body, and therefore for its possible eternity.

There are two striking aspects to this story. The first, obviously, is the gap between what is made of this body and its biological components and what we imagine as the acceptable limits to modifications of physical integrity. Embalming is not a new practice – far from it – but even so… One might be tempted to share the suspicions of the Duma deputy: are there not limits to the transformations that we put a body through? Can we treat human organs like parts of a machine that can be replaced without scruple? In other words, can we really treat a human body as one example among others of the paradox of Theseus' ship, which has become the cliché of thought experiments on identity? This paradox questions the permanence of a ship that, over the course of a long journey, has seen each of its original planks replaced by a new one. Can Lenin's kidneys, liver, nose and hands be treated like planks? The answer, actually, is anything but obvious. Above all, this answer had to be invented as the preservation of the body gradually moved towards a form of permanence. This is the second lesson of this story. Neither the Communist Party officials nor the scientists who examined Lenin's body found themselves confronted with a puzzle that they could have resolved by providing an unambiguous answer to a binary question. On the contrary, it is through responding to highly concrete and sometimes frankly trivial problems that a certain version of Lenin's body could be established, one deemed valid and adequate, as it were – a version that defined a consensus combining a physiological integrity to which everyone was attached[31] and the possibility, both political and scientific, of a form of permanence. Not much of this version could be decided in advance. 'What mattered' in this body so that it remained Lenin's body was the subject of discussions, arising from sometimes insoluble problems, and then, little by little, from agreements eventually set up as principles.

Very particular aspects were established as criteria of the integrity and thus the effective preservation of a body intended to incarnate not simply Lenin, but Leninism, as the Party of the time wished to display it: the skin colour, weight, and even, more surprisingly, the flexibility of certain joints and the elasticity of the skin – two elements considered crucial but whose permanence raises complex issues. Each of these dimensions gave rise to both ontological and technical explorations, through which the properties of the body to be maintained and the methods of its maintenance were agreed on together. This is how a large number of internal organs, which proved easy to consider as 'accessories', were removed; but at the same time it was necessary to find ways to add to the body certain elements intended to ensure the sustainability of certain properties: a series of fluids that were regularly replaced to irrigate the skin, and a mixture of glycerine, paraffin and carotene that was created to replace certain solid lipids which liquefied too quickly and caused certain parts of the body to lose their original shape.

We could easily mock these improbable mechanisms and consider them as a form of exoticism tinged with a cheap culturalism; we could describe the whole enterprise as absurd and decide in our turn, with a shrug, that in reality nothing of the real Lenin is to be found in the hybrid body – almost a cyborg – installed on Red Square. But that would be to forget that we are here to learn about maintenance and those who practise it. They are the ones who guide us on the circuitous paths that people sometimes have to follow to make things last. And it is by recognizing the ontological implications of the problems raised by the potential eternalization of Lenin's body, the inquiries in which those involved immersed themselves, and the solutions they came up with, that we can discover unsuspected dimensions of permanence. And this case shows that permanence puts to the test the simplistic (and very Eurocentric) idea of an authenticity reducible to biological integrity. It is indeed Lenin's body that has been maintained for all these years – 'cultivated', Yurchak writes nicely. A body that remains whole only if it can exist partly in pieces; a body that is actively decomposed and recomposed, and thus escapes each day from another – and irreversible – decomposition.

That being said, we must not dismiss too quickly the question of the moral positions that attack claims to permanence, and the ontological as well as economic costs that they represent. For one thing is certain: this maintenance involves a certain immodesty. This is what distinguishes it so clearly from the well-behaved horizon of duration arranged as a simple prolongation. This immodesty concerns both the thing maintained, whose properties are calibrated, and time itself, which is no longer a unique vector of irremediable and universal transformations; it also involves the means allocated to the incessant dance that conditions the performance of stability. Each of these three aspects can be subject to stringent criticism. The money and human resources devoted to the maintenance of Paris metro signage or the Eiffel Tower may be considered far too great in terms of other priorities. And the criteria used to define the integrity of Lenin's body can be called into question.

In *The Dark Abyss of Time*, Laurent Olivier gives an example of the criticisms that can be directed at the temporal horizon of this form of maintenance by focusing on the village of Oradour-sur-Glane and the particular form of preservation to which it has been subjected.[32] The case offers a perfect complement to those we have already discussed. From the end of the Second World War, the site was involved in a vast preservation operation, intended to make each material element of the place an object of memory of the terrible day of 10 June 1944, when an SS battalion massacred almost the entire population of the town before burning it down. The project went well beyond the traditional production of a 'place of memory'. In an explicit reference to Pompeii, its organizers decided to literally freeze the site, to make fixity the operator of commemoration of a life that had ended in the summer of 1944. As one can imagine, this decision entailed many complications.[33] Firstly, it was necessary to establish the initial conditions of an 'arrested life' that had not, in fact, been completely frozen by the Nazi operation. To make the village a faithful witness to what *had* happened in the past, it was necessary to ensure that nothing more should ever happen there: closing the road that crossed it, walling up the buildings still in use and offering the residents who still owned houses and fields another Oradour, built nearby. This 'cauterization', starting in 1946, was not enough. This is because, even without humans, everything within the village walls continued to live, or to die, depending on how one views things: the wooden and earthen parts of the buildings quickly disintegrated, while shrubs and weeds spread over the site. Consolidation on one side and gardening on the other were necessary: people had to take care of the place, to ensure that it remained the same. As in the case of Lenin's body (and even if, here, the horizon of permanence was immediately identified as a guideline for maintenance), the people in charge of making Oradour last forever were confronted with both dimensions that we mentioned above: the need to organize a continuous, almost daily activity of care, and the obligation to adopt over time a modular definition of the village which made it acceptable to replace certain components by others in the name of preserving a larger identity. Olivier deplores this situation, which he takes as an example of the failure of any initiative aimed at producing the conditions for an 'archaeology of the present'. As the difficulty of conservation interventions demonstrates, entrusting material objects with the mission of transmitting a memory to the future is a chimera; one that mirrors the way the relationship that archaeologists have with the remains that they discover buried in the earth or under water is itself imbued with an immense naivety.

Here are the terms Olivier uses to demonstrate his argument:

The problems associated with preserving contemporary sites teach us something not easily understood about 'classic' archaeological sites that to our mind belong to periods of history that have ended: Regardless of what we do, the past, as material creation, continues to exist and to evolve in the present. [...] We fail to see that this is not the time of the artifacts themselves, and that in our desire to maintain these remains 'in their past,' we are actually encumbering them with additions that are foreign to them.[34]

The basis of this critique recalls certain arguments of Michel de Certeau, who claims that the writing of history always has a generative character.[35] But when, in the same passage, Olivier mentions 'additions that are foreign to them', we sense a dismissive attitude based on the hypothesis that the objects preserved are inauthentic. The 'real' matter of things can only age and mutate. Everything that contributes to a negation of this transformation is a lie or an illusion. But in fact, Olivier is criticizing not so much the objects preserved by this form of maintenance, as the temporality that such maintenance claims to produce: we cannot escape the natural time of matter, that of permanent mutation. People – and above all archaeologists – must not claim to access through things a frozen past, understood as a distant point on the uninterrupted line of 'historical' time. This position is reminiscent of the heated debates that have plagued the preservation and conservation professions for decades, if not centuries – we will have the opportunity to return to them in our next chapter. Let's note for now that such a statement underlines the self-assertive and immodest dimension of a maintenance oriented towards the production of a permanence that strives to tear itself away from the inexorable time of the physical deterioration of things. Rather than criticizing the unrealistic, even untrue, nature of this gesture, the example of the village of Oradour-sur-Glane helps us to highlight everything that this maintenance involves in practice, the difficulties it raises, the negotiations with the materials it implies, and the work on which it is based.

Olivier's critical position ultimately points to an aspect that we have not mentioned so far. What he seems to reproach the preservationists for is perhaps less that they themselves believe in the possibility of a fixed material time than that they pretend 'as if' – as if one could access the past without any other intermediary than the material aspect of

THE CARE OF THINGS

the objects, as if these objects could travel through time without being transformed and without external intervention. As if they weren't *things*. In this criticism of acting 'as if', there is a potential argument lurking, traces of which can also be found in the work of Isabelle Stengers and Bruno Latour on the modern sciences.[36] Seen from this angle, this form of maintenance, in addition to its immodesty, shares with most scientists a tendency to mask some of the conditions whereby it generates its portion of reality, erasing the traces of its action and its accommodation with things and spotlighting only objects that last. For permanence to function in the three cases on which we have focused, it seems essential that maintenance should not be unduly emphasized. It is undoubtedly in connection with Lenin's body that this logic is most obvious: everything relating to the machinery of its preservation, whose heavy significance we have noted, remains secret, known only to a few initiates. On this point, the metaphor of dance as performance reveals its limits, since the complex choreographies which ensure the fixity of the metro signage, the remains of Lenin and of Oradour-sur-Glane are deployed in times and spaces inaccessible to the general public.[37] Public performance consists not so much in the interventions and their repetition as in their result: the immobile thing, present to the eyes of all for eternity. Permanence therefore seems to go hand in hand with the division, on either side of the thing maintained, between the backstage where a swarm of maintainers are busy making themselves more or less invisible, and an audience summoned to admire or, sometimes, simply to contemplate as a given an eternity apprehended directly from the material properties of objects themselves.

However, let us not quickly reduce all of this maintenance to a theatrical logic in which the thing maintained 'forever' finds itself frozen at the border which separates an innocent audience from an overcrowded backstage. True, the image seems to describe a modern version of the fabrication of permanence, one going hand in hand with a significant degree of control of the environment in which the thing exists and from which it is even partially detached. But we find in Japan a case of the maintenance of permanence that is organized completely differently, even though it testifies to an ancestral practice of heritage preservation. This concerns the maintenance of Shinto temples in the Shikinen Sengu tradition. The best-known site where this tradition is cultivated is located in Ise, about 100 kilometres south of Kyoto. The various temples and altars that comprise it have been the subject of scrupulous preservation for more than thirteen centuries.[38] The Sengu practice seems to have been very widespread until the Middle Ages, and although it is much more marginal today, it remains very much alive. The tradition is therefore literally ancestral. While we obviously cannot claim to master its subtleties, the general principle on which its organization is based seems to us to be particularly useful for understanding the forms that the quest for permanence can take. The maintenance carried out by the Sengu consists of rebuilding at regular intervals (in this case, in Ise, every twenty years) all of the sacred buildings of the complex, as well as tending to the approximately 1,600 treasures that they house. During each cycle, preservation is ensured by a complete renewal, which ends with the destruction of the old buildings. There is an enormous amount to be said about this system, which produces eternity in a movement of radical regeneration in comparison with Western definitions of architectural authenticity, and the interpretations of historians and anthropologists vary. It seems that there is at play in the Sengu cycle not only a religious activity which consists of ensuring that the spirits have the possibility of having a habitat with reliably pure and solid materials,[39] but also the transmission of an ancient know-how to each generation of exceptional artisans who complete their training by participating in the event.

In terms of the questions that concern us, we find in this maintenance the two main characteristics identified so far with regard to the manufacture of permanence. Their features are even here clearly accentuated. Modularity, evidently, is at a maximum: it is a generalized replacement that is the main driver of permanence. No one in this situation would see the slightest paradox in the story of Theseus' ship – quite the contrary. It would probably be only once all of its planks had been replaced that one would consider that the ship in question had actually been preserved. The choreographic dimension also seems to be taken to an apogee. The dance of maintenance is ritualized to the extreme, and spreads out over seven to eight years of activities regulated down to the smallest detail, from the cutting of the cypresses cultivated in the surrounding area – carried out several years before the start of the actual construction work – to the burial of the ancient versions of the treasures, via the manufacture of fabrics, swords and pottery. It involves a considerable number of participants who flock from all over the country so that the spirits can be successfully moved to their refurbished refuge and live there peacefully for the next twenty years.

But this very particular form of maintenance, where architecture and craftsmanship mingle, encourages us to nuance our understanding of the fabrication of permanence. In particular, it seems to be part of a much more modest mentality than the cases we have looked at so far. Such an assertion may seem surprising given the gigantic resources involved during each cycle of complete reconstruction. However, as the rites in question bear witness, each operation of maintenance cultivates, in the Shinto tradition of which the Shikinen Sengu is itself one of the rituals, a relationship with life far removed from the almost arrogant position characterizing the situations described earlier. Rather than scientists or technicians seeking to control the behaviour of the thing they are taking care of and the vagaries of its environment, maintainers here are akin to 'couriers', respecting as much as possible the ecology of the environment in and with which they work, to produce a form of continuity that is both material and spiritual. Furthermore, the disassembly and reassembly activities specific to the Shikinen Sengu are not kept away from the public. If some rites are reserved for a small community of initiates, most of the operations which span the eight years of work are marked by sometimes spectacular public ceremonies. The human presence in the manufacture of permanence is therefore not erased – quite the opposite. In the same way as the tools and the different materials used, the craftspeople maintain a recognized place in the process of carrying out an eternity which makes no attempt to persuade as many people as possible that it is due only to the intrinsic properties of the thing that lasts.

To conclude our exploration of the fabrication of permanence, we must once again insist on the ontological issues of this form of maintenance, from metro signage to the temples of Ise, including Lenin's body and the village of Oradour-sur-Glane. While what is at stake in the often improvised practices of prolongation is quite vague, and rarely articulated around clearly identified precepts, the configurations that tend towards a horizon of eternity invest explicitly in principles that end up establishing themselves as a doctrine. These principles all point to the visual dimensions of the thing. The shape appears to be the essential characteristic that must be inscribed in the time of permanence. It is the exterior aspect of things that needs to be made to last 'forever', their surface. Of course, the relevant features of this physiognomy as a formal aspect are, as we have seen, subject to discussions, debates and even collective inquiries on the part of those who take care of things. But it is indeed in this capacity, and in a certain way at this 'price', that things are maintained so as to have a potential permanence. They last through their appearance, which is made durable on a daily basis.

Slowing down

Even if it has little to do with a certain overenthusiastic improvisation found in the apparent ontological relaxation of prolongation, the fabrication of permanence is based on material constraints with which it is possible to compromise. There is nothing obvious about this, of course. Permanence involves colossal work, and the very possibility of eternity is at stake every day depending on the means dedicated to it. Despite everything, it is linked to a sort of ontology of compromise which accepts that the thing maintained can undergo a certain number of transformations deemed acceptable. Such a position seems absolutely unimaginable in other situations. There are worlds in which certain objects are almost untouchable; worlds where the kind of material developments that we were able to discover by following the trajectory of Lenin's body or that of the temples of Ise would constitute an unforgivable transgression, since the modifications made would bring about an irreversible ontological break. While it does not allow things to settle into eternity, the time of maintenance specific to these configurations is not the time of an anodyne and almost casual prolongation that merely projects onto the horizon a series of tomorrows, one day conquered after another. Profoundly unsettled, this time combines the certainty of inexorable decomposition with the fierce desire to slow down the pace of deterioration.

This form of temporality is not common. It is mainly found at work in institutions such as archives, libraries or museums, responsible for taking care of unique things whose long-term sustainability they guarantee. During an investigation carried out in the Austrian National Library, Moritz Fürst described some interesting aspects of this.[40] By discovering the way in which certain employees of this library worked, he was struck to see that even though books seem to be understood in most departments of the institution in terms of their cultural and historical 'contents', they are handled in the conservation service simply as objects, of which only the material properties matter. Far from any linguistic and discursive consideration, it is these material properties that are explored here, examined with the aim of finding signs of deterioration. Cracks, peelings, mould: as objects, books are carefully examined for any fragility. But the particularity of this fragility is that it is captured in a biographical trajectory. Books live, and therefore age. The marks of transformation that we discover in them are traces of this life and symptoms of this senescence. Faced with the ageing of books, explicitly seen as inevitable, library curators allow themselves merely to slow down its pace, or at least prevent it from accelerating. This requires daily attention, encouraged by the circulation that lies at the heart of the very organization of the library. Each loan is an opportunity for a quick inspection, which allows the curators, in a sort of airlock between the reserves and the lending room, to check that the book is still in good enough condition to be released and, if necessary, to make small-scale interventions before it leaves, or even to interrupt the process in order to make it undergo emergency treatment, before it departs into the hands of a reader. This occasional monitoring is obviously very uneven, since it only benefits the books that move most often. So there are also regular inventory campaigns that focus on a sub-part of the collection. Much more rarely, a big general inspection covering the entire collection is carried out.

Fürst was lucky enough to witness a campaign of this nature while he was looking into a related operation, which was in some ways the trigger: the digitization of the entire store. This gigantic project represented an exceptional opportunity to assess the state of ageing of all the books, since they all had to be moved to the scanners of the service providers responsible for the operation. It was enough to require that they all pass through the conservation department to ensure that each book was checked. But the programme also increased the potential wear and tear, with the handling required for digitization representing a significant risk of suddenly increasing the deterioration of certain books. The very process of digitization – an incongruous event in the books' lives – was thus designed with the aim of avoiding this sudden acceleration. There was no question of subjecting each copy to the torture that we have all probably practised once in our lives, and that consists of breaking the spine of a book in order to ensure that two opposite pages are flat and that the texts and images appearing there are digitized (or photocopied, if we go back a few years) without their visibility or readability being affected. In this kind of large-scale digitization, involving old and rare collections, service providers use a dedicated system aimed at making the books subject to the most respectful handling, similar to that required for reading. Installed on this particular scanner, the book is only opened to an angle of 120 degrees maximum, and a human operator turns the pages. The issue, as the curators explained to Fürst, is that the manipulation represents the equivalent of an 'additional occurrence of reading', no more and no less: an ordinary episode in the life of the book, one that will not hasten its ageing. Interestingly, the inspection prior to each batch sent for scanning is the exact counterpart of this system. The curators do not allow themselves to manoeuvre the books in all directions under the pretext of evaluating their condition. They are content to hold them in the same conventional reading position as that which the scanner will subsequently adopt, thus ensuring each time that the life of the books in their charge remains as peaceful as possible.

As for possible interventions to which the works are subject when worrying signs of deterioration are found, they remain minimal, in most cases. From adding glue dots to sometimes delicate removal of mould, and reinforcing the binding, they are never considered as 'repairs' strictly speaking. In the vocabulary of the curators, each operation thus aims to temporarily 'secure' the state of a book: that is, to prevent any sudden deterioration. The relationship with time that is cultivated here therefore does not involve a split. Books in their materiality do not escape their destiny: that 'steady movement towards decay',[41] considered by everyone in the library department as the common condition of the objects of maintenance.

After this short visit to the conservation department of the Austrian National Library, we come away with a first idea of maintenance which is organized around slowdown, and we can start to understand how it differs from ideas of prolongation and permanence. But to appreciate the subtleties of this very particular relationship with time, and its practical and ontological implications, we must go where these issues are even more pressing. To do so, there is probably no better place to start than the backstage of a large museum. It is these implications that Fernando Domínguez Rubio demonstrates, who wrote a fascinating book on conservation at the Museum of Modern Art in New York[42] a few years after studying the case of the *Mona Lisa*.[43] It is with the latter that we will continue our journey. Just as Lenin helped us understand certain unsuspected aspects of the pursuit of permanence, *Mona Lisa* has much to teach us about maintenance through slowdown.

As a good guide, Domínguez Rubio first encourages us to detach ourselves from our habits as spectators. If we want to see and understand works of art as they are treated for their conservation and therefore their maintenance, we absolutely need to take a step aside and extricate ourselves from the frontal, distant, immobile gaze which has become, over the centuries, the main condition for a successful experience of the modern museum. The step aside, in this case, is also a step forward. Come closer, Domínguez Rubio tells us, and by scrutinizing the works, looking behind them and to the sides, forget art for a time, forget meanings and mental representations, and take stock of the material heterogeneity and the fragility of the objects you're looking at. Of course, due to this very fragility, it's impossible for us to make this move ourselves each time we enter a museum. In practice, it's not so easy to break away from the refined and platonic face-to-face encounter that museums and galleries set up. But Domínguez Rubio does it for us, following the hard-working crowd of people responsible for taking care of the works on display, observing the movements of their hands, their tools and their own gaze. What do we see of the *Mona Lisa* by adopting this position? First, a composite assembly: multiple layers of paint and varnish, applied to a very thin poplar panel, all enclosed in two framing systems. We then understand that this assemblage is subject to permanent movement, far from the impression of inertia that even an attentive visit to the Louvre produces. As poplar wood is very sensitive to changes in temperature and humidity, it is constantly 'at work'. This movement puts paint and varnishes to the test. It causes the numerous cracks that can be seen when examining the surface of the painting. It would undoubtedly be unfair to blame the poplar alone for this, though. It seems that a large number of cracks appeared very early on in the life of the *Mona Lisa*, following the use of a new mixture that Leonardo da Vinci had made in order to make his oil dry more quickly, and which gave insufficient support to the internal movements of its surface of application. More worrying than these cracks, a large split measuring eleven centimetres appeared, probably at the beginning of the nineteenth century, just above the head of the *Mona Lisa*, and threatening to dislocate the painting if it widened. The colours have changed too. Some have completely disappeared from the picture, which no longer presents the brilliance of yesteryear. *Mona Lisa* is today devoid of the reds which paid a vibrant homage to her youth when the master painted her. The austere appearance of the painting as we have known it for several decades is the result of these mutations.

We can see in these different transformations the consequence of the tribulations of a busy life. After all, it is completely to be expected that after several centuries of existence, an object, whatever it may be, will show signs of time. Indeed, isn't this partly how we can appreciate its authenticity? However, in the eyes of those who play a part in the conservation of the *Mona Lisa*, these deteriorations are much more serious. They represent direct threats and a proof that the painting will ultimately disappear. The danger may seem very remote on the scale of a human life and yet it is very evident and constantly present to the minds of the people who take care of the painting. To illustrate and clarify this, Domínguez Rubio explains that, rather than disappearing in the strict sense, that is to say without leaving a trace, objects preserved in museums are destined to become over the course of time 'something else', like an ice cube that eventually spreads out into a puddle of water which itself then slowly evaporates into the ambient air. There is nothing specious in this metaphor: conservators all have specific examples in mind that justify their fears. For example, the red chalk lines of Leonardo da Vinci's famous self-portrait are almost invisible today. Soon, all that will remain is a sheet of paper five centuries old which can no longer be said to be the 'portrait of an elderly man', or – and this is even more unfortunate – a work by the master.

All conservation work therefore consists of confronting these threats by striving to slow down the great march of disintegration. And if each museum is in itself a vast slowing-down machine, whose backstage is teeming with operations dedicated to this infinite task, the case of the *Mona Lisa*, that supreme masterpiece, emphasizes the significance of this process in detail. How is the process of deterioration slowed down? In recent decades, a very large portion of the resources devoted to it have consisted of containment techniques. In order for the painting to age as slowly as possible, the atmosphere around it is obsessively controlled. The principle is not exceptional in itself: it is the characteristic of museums to provide a healthy and stable climate for the works they house. Three dimensions usually characterize this climate: temperature, humidity level and light. Each material from which a work is made – or rather, each combination of materials – requires a specific balance between these three components. Domínguez Rubio tells us that oil paintings, such as the *Mona Lisa*, do not tolerate variations well. The ideal temperature for them is between 18 and 24 °C, while the humidity level should remain between 40% and 55%. The light to which they are exposed must not exceed 200 lux, with ultraviolet radiation less than 80%. To these three dimensions, we must add the question of the composition of the air itself. Paints are sensitive to a large number of pollutants and even to oxygen, which contributes to the deterioration of colours.

The great difficulty that the production of this type of climate represents lies, as one suspects, in the need to focus on the other side of museum activity, namely exhibiting the works to the public. Ensuring that the temperature and humidity levels remain stable throughout the season with thousands of visitors per day is a real challenge. Even more problematic, the very presence of these visitors affects air quality and accelerates the deterioration of the paint. For most works, the conciliation between preservation and exhibition simply involves long stays in reserves, spaces with a much more easily controlled atmosphere in which the possibility of leaving them in almost complete darkness is an invaluable advantage of slowing things down. The *Mona Lisa* cannot benefit from this alternating existence, in which the painting can be exhibited in exceptional circumstances and the months hidden from view can provide a well-deserved rest. Leonardo's painting is not allowed any period of isolation. To ensure that it does not suffer too much from this uninterrupted public display, considerable resources have been deployed. Thus, in 1998, Louvre officials decided that, given the ever-growing success of the painting, the Grande Galerie where it had been exhibited until then could no longer guarantee its security. The space to accommodate the spectators had become insufficient and the control of the temperature and light in the room was not satisfactory. They therefore decided to move the *Mona Lisa* to the Salle des États, which first had to be completely transformed so that it could provide the best possible climate. Five years and nearly 6 million euros were necessary to set up these surroundings. A system for 'purifying' the air was installed, as well as a complex device capable of dynamically stabilizing the temperature and humidity of the room depending on the season and the number of visitors. Lighting, in particular, was the subject of significant technical investment in order to produce a mixture of artificial light and natural light, also capable of being adjusted throughout the day and over the seasons. However, the room's fittings are only part of the slowing-down system put in place to install the *Mona Lisa* in exhibition conditions deemed acceptable. The room was lined with a caisson whose unique character is both the symptom of the exceptional nature of Leonardo da Vinci's painting and a condition of its success. This caisson duplicates most of the functionalities of the room's fittings. It is equipped with a dual-flow air-conditioning system which further stabilizes the temperature, re-filters the air in the room and generates the purified atmosphere suitable for the painting. More surprisingly, this move also provided an opportunity to add a lighting system for the painting using LEDs, intended to further reinforce the impression of natural light, but also to bring about, through the play of the shades of colour used, a sort of passive 'restoration' of some of the faded hues.

So much for this kind of lockdown, a hybrid method combining the requirement to slow down with the need for exhibition. Although spectacular, such a measure represents only a small part of the very long story of what those in charge of the painting's welfare did to ensure that it did not deteriorate too quickly, but could age as slowly as possible. It is difficult to gain any precise idea of the details of the care deployed throughout its five centuries of existence. We know a few of the measures taken, which highlight the difficulty of the exercise, the variability of the requirements and the problems and ambiguity of any form of maintenance. For instance, during the first half of the twentieth century, to compensate for the instability of the paint, several layers of varnish were applied. However, it is this varnish that is today considered responsible for the dulling of colours. An innovative and ambitious stabilizing material, it has actually accelerated material deterioration. Furthermore, the eleven-centimetre split, a major source of concern, was the subject of several interventions. During the nineteenth century, a wooden crosspiece was inserted into the back of the poplar panel to prevent the split from spreading further. Glue was even applied to the crack, thereby aggravating its fragility by making the whole thing even more brittle. Finally, in 1951, a flexible oak frame was added to the initial framework in order to ensure the stability of the whole while allowing the wood to 'work' in acceptable proportions. Then there were the countless monitoring activities aimed to identify as quickly and as closely as possible the slightest signs of acceleration in the work's deterioration. The painting has been the subject of a myriad of scientific tests. In 1974, an annual visit was established in order to precisely measure the slightest material change. Since 2007, monitoring has been further stepped up: the *Mona Lisa* is now equipped with a permanent device made up of nine sensors placed behind the painting and which, every thirty minutes, transmits a set of data on the painting's behaviour via Bluetooth.

Ultimately, there is a certain symmetry between this configuration, where slowing down involves a surplus of equipment, and the frantic dances dedicated to producing permanence. In both cases, there is a great contrast between the time of the thing and that of its maintainers. On the one hand, it is daily agitation which produces – and constantly reproduces – an eternal fixity; on the other, it is the acceleration of monitoring which ensures optimal conditions for slowing down. But the similarities end there. One absolutely crucial difference remains: the thing is maintained, but the horizon of its future, inevitable and certain disappearance, is forever in sight. There is no question of escaping it. The thing can do nothing but wear out, irreparably. And it is therefore in these terms that time poses a problem here: a vector of deterioration from which it is unthinkable to completely escape. That said, while it is impossible in this situation to escape from time understood as a universal condition of the inexorable transformation of matter, as we have seen, maintenance is not necessarily thereby reduced to inaction. It is a cautious process, though. Making things last goes with a form of domestication of this inexorable time, a subtle art requiring a light touch which, while it allows room for manoeuvre, must nevertheless come to terms with extremely strong material constraints. Maintenance here is similar to resistance in the mechanical sense of the term: a kind of braking. But it's a braking which must avoid derailing anything.

The difference also plays out in ontological terms. This deterioration, which presents itself as an irrefutable datum from which nothing escapes, corresponds to a very precise definition of the thing to be maintained, whose very existence is conditioned by a certain idea of integrity and originality. There is no modularity here. The thing can last only if it remains whole and unique. These two closely linked dimensions drastically limit any intervention, and place enormous responsibility on each maintenance operation. If this goes wrong, it could either accelerate the process of deterioration instead of slowing it down, or trigger an outright disintegration of the object of care. The form, the appearance and even the surface, which we saw were the keys to the possibility of permanence, are not enough in this case to maintain the thing whose list of essential components, in the literal sense of the term, is much longer. One cannot replace the missing colours of the *Mona Lisa*, any more than the entire binding of an old book can be restored. By making small adjustments to the thing itself, and sometimes by making major adjustments to its environment, one can simply ensure – but this is far from easy – that it can age as slowly as possible.

However, we must be careful not to imagine that these ontological requirements are reducible to clearly established criteria that just need to be followed blindly in order to preserve authenticity. Here too, concrete practices of maintenance involve trial and error, inquiries and tinkering. They sometimes give rise to lively debates and real dilemmas.[44] The criteria themselves are not fixed in time; they evolve with changing historical concerns and technological innovations. In this respect, things are not very different from what we have been able to understand about the permanent form of maintenance. As rigid and monitored as it may seem, the care of things, once we look at the specific conditions of its accomplishment, is a fundamentally exploratory affair.

With these clarifications in mind, let's dwell a little longer on the manufacture of time associated with the idea of slowing down. As we discovered, in particular from the adventures of the *Mona Lisa*, we must admit that it involves a quite exceptional process. Objects to which we devote so many resources in order to treat them strictly in the mode of linear slowing down are rare. Usually, in the field of art, museums and their reserves provide a shared infrastructure, capable of ensuring common air conditioning for all the works. More recently, we have seen gigantic warehouses appear around the world, called freeports (with all the attendant connotations), established in cities with favourable tax conditions. In these literally 'separate' places, like ivory towers, are housed works that can change ownership on the art market without having to face the hazards of international transport, or even the trial of exhibition. Nevertheless, in these spaces, each object is housed in approximately the same way, and above all the slowing down is almost exclusively carried out through the control of atmospheric conditions. No intervention is carried out on the works themselves. It is the 'lockdown' model that is best suited to maintenance by slowing down.

On the other hand, the field of art includes another type of operation that does indeed consist of intervening on the works themselves: all those activities grouped under the term 'restoration'. How are we to think about these practices which, as we have seen in the case of old books from the Austrian National Library, involve retouching, sometimes adding or removing material? How long do they last? We might be tempted to associate them with the issues of permanence, but that would probably be going a little too quickly. There is indeed something in restoration which links it, despite its interventionist character, to the gesture of slowing down. Let's go back to the concerns of Untergunther, the team of clandestine restorers with whom we began this chapter. We saw that its members had intervened on the clock of the Panthéon at the last minute. Oxidation, they noted, was on the verge of literally making certain parts of the mechanism disappear and, with them, the original clock: 'A restoration beyond this period would in fact consist of remaking almost all of the parts. This would no longer make it a restored clock, but a facsimile.'[45] Their ambitious operation, however, did not have the objective of definitively avoiding the disappearance of the clock. By cleaning and repairing damaged parts, the collective never claimed to freeze the object of its care once and for all. As we have already mentioned, everyone knew full well that the process of deterioration would resume the minute the restored clock was put back into position. Restoration remains deeply anchored in this logic of universal deterioration, one that traces a single timeline. On the other hand, the slowing down that it achieves is of a different order from that which we were able to observe with the *Mona Lisa*. It goes through what we could call a retro effect. It is neither a simple braking nor a change of trajectory that is at stake, but a step back on the same timeline, a line whose uniqueness guarantees that the restored thing remains the same. Without achieving the claim to permanence, the gesture is much less modest than that of the meticulous domestication of the conditions of ageing. The slowing-down effect is in fact greatly amplified, since here one can go into reverse: it's almost a process of rejuvenation. But it remains no less delicate. The botched restoration, now world-famous, of the *Ecce Homo* in Borja – a painting from the beginning of the twentieth century in the church of a village in Zaragoza, which an amateur set about retouching in 2012[46] – testifies to this, as do the countless long-term controversies that plague the profession.[47] The retouching that allows something to go back in time while preserving its integrity is an extremely complex matter. Fighting too intensely against one form of decomposition sometimes leads to triggering another. By wanting to brake too hard, and – even worse – by backing away too quickly, we risk seeing the object of our care slip through our fingers, and its crucial authenticity evaporate, while by derailing it from the timeline that we sought to assign to it, it becomes something else.

Stubbornness

Let's continue along our temporal journey, while keeping in mind the need for openness to variation. As a final step, we propose to venture into the outskirts of maintenance, just to discover a few situations that will detach us once and for all from the simplistic image of a voluntary human enterprise, entirely dedicated to the mastery of matter and time. While the expression 'to make things last', which we have used extensively until now, is attractive, it can in fact lead us down the wrong path and make us miss the subtleties of the activities and problems that run through the care of things. The different situations described in this chapter have already amply shown that, contrary to what a first impression might perhaps suggest, the care of things cannot be reduced to a banal matter of human control over objects and their environment. None of the different forms of intervention that we encountered is organized in accordance with a completely unilateral relationship in which women and men shape the time of things as they wish. Humans grope, hesitate, explore, improvise… And above all, as we began to see in the previous chapter, things themselves have their own say, and their own gestures to make. They resist, object, evade; people must constantly negotiate with them in a dance in which the conditions of a common becoming are woven, gesture after gesture. However, despite the attachment to this vocabulary of material diplomacy, there is still a great temptation to identify the source of maintenance, and therefore the driving force of reorganized temporalities, in human actions alone. This is probably due to the fact that the most striking situations, those which most easily occupy our mind when we imagine what 'maintaining' means, are still those that relate, if not to the heroic act, at least to the presumptuous movement – that of use extended, of time saved, of memory preserved. It is therefore worth exploring the vast continent of maintenance even further to identify borderline cases, or even experimental forms where things last well beyond any human prediction or will. These situations will help us to get rid of the dualistic reflex that separates in advance the actions of things from the actions of humans.

It is precisely this type of disruption that geographer Caitlin DeSilvey seeks to trigger in her book, whose title alone (*Curated Decay*)[48] probably sounds like an unbearable provocation to the ears of conservation professionals. Based on a few examples from the field of heritage conservation, DeSilvey attempts to identify, in a deliberately speculative way, alternative forms of preservation to those that pertain exclusively to the model of the mastery of matter. Can we imagine, she asks, ways of taking care of things that do not boil down to a head-on fight against material deterioration? Starting from her disturbing experience as a curator tackling artefacts found in a nineteenth-century property in Montana, her investigations take her to various places: an industrial site in the Ruhr, a port in Cornwall, and Saint Peter's seminary in Glasgow. That icon of brutalist architecture was built in 1966 and abandoned twenty years later, before its future became the subject of passionate debate at the Venice Architecture Biennale in 2010. Each time, DeSilvey looks at 'experimental' preservation projects and the sometimes heated discussions they generate. In doing so, she sketches out a conservation model that she calls 'post-preservationist', describing its fundamental principles, but also its points of tension and lines of flight.

At the heart of this project, she explains, lies the question of the connections that humanity is capable of weaving between memory and matter and, more precisely, the possibility of moving beyond a modernist framework entirely focused on the idea of immutability. Can we disconnect the memorial role given to certain objects from the quest for material fixity? It is therefore ultimately less the question of mastery and control that this form of alternative maintenance raises than that of the relationship to change. Explicitly arguing in favour of such shifts, DeSilvey spreads her net even wider: she seeks 'ways to inhabit change rather than deny or deflect it'.[49] This position ultimately amounts to adopting a diplomatic and benevolent, even frankly enthusiastic, view of deterioration and entropy. As she explains, instead of seeing these processes as threats, it is possible to consider them in a positive or, at the very least, more balanced way. Rather than being synonyms for disorder and loss, a post-preservationist perspective encourages us to get closer to their initial status in science, where they are used to describe the capacity of matter to transform itself; deterioration and entropy foster attention to the multiplicity of material *agencements*, in line with the work of Karen Barad, Janet Bennett and Tim Ingold.[50]

The initiatives described by DeSilvey appear as so many attempts to positively introduce a certain dose of laissez-faire, and even of letting go or neglect, into the exercise of heritage conservation: a way of accepting the material transformation of things. By asking that the management of a ruined industrial site be organized in accordance with a 'principle of non-intervention', by accepting the imminent disappearance of a lighthouse or a port threatened by storms and rising waters, by imagining how to treat a building 'like a landscape', the institutions concerned each engage in their own way in a productive relationship with deterioration. It is remarkable that, each time, it is not just a link to the things maintained which is at stake, but a rethought relationship with their environment and more generally with 'nature'. Animals, plants, seas... Laissez-faire makes possible the participation of new entities in a thing's life – the latter we can now accept is no longer a matter of human responsibility alone. We can even detect in some of these initiatives the trace of another form of preservationist ideal: no longer the maintenance of objects and buildings placed under the attentive control of women and men, but the maintenance of a nature whose conservation could be ensured only if it were completely protected from human intervention. In certain people that

DeSilvey meets there is a temptation of the wilderness, a fantasy of the landscape quite untouched by humanity,[51] a desire to completely leave room for 'nature' in a gesture of withdrawal which ultimately transfers responsibility for the time of the thing that had hitherto been preserved and leaves human beings as mere spectators, free to be newly affected by the spectacle of deterioration. But this temptation is never fully realized, as DeSilvey explains. While it appears in certain programmatic texts, and in alternative doctrines, it does not pass the test of practice. For a whole host of reasons, first and foremost personal safety, it turns out to be impossible to completely let 'the elements' do their work without ever intervening again. On the contrary, it is the invention of delicate partnerships that DeSilvey encourages, in which the place of women and men continues to oscillate between 'too much' and 'not enough'. By giving a positive value to deterioration, it is not only a question of recognizing the permanent recomposition at work in the most subtle combinations of matter, but also of constantly adjusting the place of human intervention and finding the concrete conditions for carrying out the 'positive passivity' that Simmel speaks of in relation to ruins.[52]

Regarding relationships with time, we can see in DeSilvey's work a half-empirical, half-speculative meditation on the domestication of loss and the work of mourning transposed to the world of objects. But this is only one aspect of the situations she describes. If we place ourselves on the side of things and their own actions, as DeSilvey regularly encourages us to do, we can also see in the post-preservationist gesture a recognition of their capacity to last in their own way, to persevere at their own pace, well beyond human intervention. It is actually because they do not disappear suddenly, once we decide to no longer conserve them according to the canons of modern heritage conservation, that we realize the extent of the perseverance of things. They sometimes seem to persist in continuing to exist, and invite human beings to take care of them, reversing the relationship of control that a narrow definition of maintenance as 'making things last' would lead us to expect.

This is the direction that the work of Marisa Cohn indicates, in a completely different context.[53] The ethnographic investigation that she conducted in a space science laboratory on the west coast of the United States provides valuable insights for understanding practices of maintenance that occur in situations where the issue is not to prolong the possible use of the object of care, nor, strictly speaking, to fight against the process of material decomposition, or even to slow down its pace. The thing Cohn focuses on is very complex. It is indeed one of the flagships of human technology: a spacecraft dedicated to the study of Saturn and its satellites – one of the most impressive ever sent into space. Launched in 1997, the spacecraft in question was meant to complete its mission in the mid-2000s, after a long journey and several years of observations. However, and this is where things get interesting for us, its life did not end as soon as had been imagined. The spacecraft continued to operate and communicate its observational data. So much so that the executive team of the laboratory in charge of it endeavoured to find the means to finance a new 'mission' so that between 2008 and 2010 one team of researchers and technicians would be officially responsible for maintaining contact with the spacecraft, ensuring its proper functioning and processing the data that it continued to transmit; then another team would take over, from 2010 to 2017, before the spacecraft ended up projecting itself into Saturn's atmosphere to self-destruct while minimizing the risks of damaging one of its moons.

If we compare it to the cases we have explored so far, the history of this spacecraft and the team responsible for taking care of it seems strikingly different. What happened at the end of the 2000s and the establishment of the first 'additional' mission has little to do with the desire to 'make last' something which threatened to disappear for one reason or another. Almost the opposite, in fact: the spacecraft was not really bothered about the technical and budgetary planning of humans. It continued to exist and function beyond the programmed limits of its 'theoretical lifespan', skilfully calculated by the small team that had designed and manufactured it and sent it into space. Even if we must obviously not exaggerate by completely reversing the modalities of action here, we can nevertheless read this case as an invitation to significantly shift the semantics of maintenance and to recognize that the spacecraft is, on this occasion, the main subject of the verb 'to last'. And, more striking still, it was the spacecraft which, by persisting in lasting, encouraged some of the people who were taking care of it from their positions on Earth to prolong their own operations, to stay in the dance. Even if they obviously had the final decision, and could have abandoned the spacecraft to its fate by letting it

THE CARE OF THINGS

wander on through the depths of space, they could not bring themselves to become detached from this technological jewel while it continued its interstellar existence, and they determined to do everything they could so that their own involvement would not end. It was, in a way, the thing that 'made last' the action of those who until then had taken care of it, not the other way around. It was the thing which insisted on extending the dance of maintenance.

This semiotic reversal is far from being a simple figure of speech. Moving from one position to another had very concrete repercussions on the way in which the scientists and technicians worked within the laboratory responsible for the spacecraft during the two complementary missions. Cohn explains how the 'extension' of the lifespan of the spacecraft and all of its components ended up generating frictions in the organization of the teams, both on the side of the researchers and on the side of the so-called 'support' occupations. Over the years, the software infrastructure that ensured communication with the machine and the processing of the data it generated had grown, with hundreds of small 'patches', additional elements of which the teams had kept barely any trace, being added. These were in fact, at the time of their implementation, considered insignificant and of no consequence for the overall balance of the thing, which in any case was not intended to last forever. Years later, this 'glueware', which had until then ensured fluidity of use and a certain short-term effectiveness, became cumbersome. Its complexity and opacity made migration to the new information system implemented at laboratory level difficult; the lab executives had also reorganized all the technical services by centralizing IT development and maintenance. The complex assemblage around the spacecraft has gradually moved away from being a finely crafted engineering mechanism of which each small element was controlled, to being treated as a geological environment within which layers had accumulated and links were woven, often invisibly, like the relationships between tree roots and certain mushrooms. Some kinds of technical knowledge had also gradually become rarer within the team. The computer language of the 'orders' sent to the spacecraft, in particular, was almost obsolete by the start of the second additional mission in 2010. It was familiar to only a handful of people, some of whom even postponed their retirement to ensure the continuity of the activity.

It is the result of a slow but inexorable bifurcation that Cohn describes. It's as if, even though they refused to let the spacecraft become completely autonomous and drift on through space, the teams responding to its calls had found themselves adrift, isolated from the rest of the laboratory. The latter, for its part, had followed a trajectory of organizational and technological innovations, the routines of which had been reinforced by successive projects, with their share of international partners, technological 'breakthroughs' and service providers of all kinds. Detaching themselves almost completely from the laboratory and its development, so as to remain attached to the spacecraft: this seemed to be the price that had to be paid. Until then the timelines had been aligned relatively unproblematically; now they split apart one by one, their divergence weighing more and more on the activities of those involved. The time of more or less irreversible obsolescence of the various kinds of on-board software, the time of the wear and tear of the spacecraft's materials, the time of the delicate transmission of scientific and technical knowledge, the organizational time of the laboratory teams, the political time of national and international space programmes, the time of the careers of scientists and technicians... So many lines of flight which ended up becoming desynchronized and contradictory until some of them separated almost completely from the rest and reorganized themselves around the time – now running out – of the persevering spacecraft.

But what about that time, precisely? How did the technical teams participate in the maintenance of this thing which ended up lasting much longer than initially envisaged? In Cohn's descriptions and analyses of the concerns of these teams and their concrete practices, we find elements reminiscent of what we discovered about the challenges of a maintenance oriented towards the slowing down of decomposition. The organization of 'additional' missions, for which it was necessary to negotiate hard budgets and resources significantly lower than the initial mission, generally highlighted the idea that the spacecraft was living its very last years. As in the case of works of art, which curators know are doomed to a tragic destiny, the spacecraft was used and maintained in the perspective of its inevitable finitude. However, the practical consequences of this shared orientation differed significantly from the idea of slowing down, since at this stage there was no question, for the people taking care of the machine, of slowing down its ageing. As Cohn shows, the main challenge for the technical teams was to carry out everything that could help the spacecraft end its days in the best possible conditions – 'to decay "gracefully"', as she elegantly puts it.[54] This resulted in a relationship with the spacecraft that was very different from that which prevailed during the first mission. Those who were responsible for its direct maintenance now took a much larger place within the team. They found themselves in a position to refuse certain manipulations demanded by scientists accustomed to getting the most out of their working instrument, because they considered them too brutal and because they risked damaging the functioning of a particular element, in the software or the material. The steering wheels, for example, presented increasingly significant behavioural anomalies, undoubtedly due to repeated use at low rotation speed, something that had not been imagined by the design team and now prevented certain operations from being initiated. Gradually, the engineers, then the scientists themselves, began to develop an emotional relationship with the spacecraft, which they spoke of as if they were describing an old person, stubborn but fragile and worthy of every attention. In this form of maintenance – which she does not hesitate to describe as 'geriatric' – Cohn shows that the technicians ultimately followed an approach quite similar to that which DeSilvey imagines in the field of heritage conservation. They strove to work *with* deterioration rather than *against* it. A large part of their activity thus consisted of getting scientists who were not initially sensitive to the issue to accept this inexorable deterioration: the spacecraft was ageing, it was worn out, diminished, they had to resign themselves to the fact – especially since this deterioration was accompanied by a significant drop in financial resources, which also entailed a certain modesty. Thus, at a meeting organized by the team's IT specialists, who were beginning to worry about the possibility of software failures due to downsizing and the abandonment of a complex access rights management system for the machines, the technical manager spent a long time making everyone aware that the question was no longer 'whether' breakdowns were going to happen, but to understand their nature and the adaptations to which they would give rise, *inevitably*. Over the course of the last mission, problems and breakdowns were no longer treated as failures or interruptions, but accepted as stages in 'a productive process of letting go'[55] – a direct echo of DeSilvey's theoretical propositions.

The parallel with what is usually called the 'end of life' for humans is striking. It is also fully accepted by Cohn, who uses it as a means of emphasizing the register of small gestures and attentions that the engineers responsible for the proper functioning of the spacecraft developed. But before returning to this comparison and what it can tell us about this form of maintenance, let's take another look at the main question we are addressing in this chapter: in what ways is time a problem here? And what kind of temporality does this maintenance help to manufacture? One word describes fairly closely what is happening, while marking the difference with the other cases we have encountered so far: stubbornness. Not so much the stubbornness of the people who insist at all costs on continuing a relationship with the thing beyond its planned end (even if that is also an issue here, of course), as the stubbornness of the thing 'itself', which persists in lasting and whose insistent presence calls on those people not to cease their action, not to undo too quickly the links which unite it with them. Not to stop dancing. Of course, the case of this stubbornness is exceptional, and there is no question of imagining that we could make it represent a vast swath of maintenance activities. But when we see it in the context of DeSilvey's reflections, its singularity has the merit of shaking up our habits of thought and forcing us to recognize what we had already begun to understand: things also have their own ways of lasting and, in the dance of maintenance, they are sometimes the driving forces.

In ontological terms, what is at stake here is just as interesting, since it is precisely the entanglement between the thing and everything that plays a part in its existence that the idea of stubbornness brings out. It is undoubtedly from this point of view that the comparison with the relationship of care found in geriatrics is the most telling. A thing's stubbornness to keep on living, like a person's, seems in fact to operate both as a test and as a revealer of sociomaterial interdependencies hitherto relegated to the background, interdependencies which constantly oscillate between support and hindrance. As we have seen, this is the value of the notion of *attachment*,[56] which underlines all the ambiguity of stubbornness, and with it the ambiguous character of care itself. If the thing demonstrates its autonomy by lasting longer than it should, it thereby actualizes that which attaches it to those people who ensure its maintenance – people on whom it imposes reciprocal *obligations*, just as caregivers are obligated to the elderly or sick people they support. Stubbornness intensifies attachments, which are always double-edged: they are a form of attention, even affection, between those cared for and their caregivers, between the things maintained and their maintainers, but they are also sources of constraint for both sides.[57]

Finally, by encouraging us to consider forms of constraint and obligation, the idea of stubbornness opens the way to the exploration of situations that at first glance we would not have seen as connected with maintenance and the care of things. This is the case of the management of nuclear waste, all those entities whose fantastic lifespan continues to pose a problem, if only because of their very uncertainty, and forces us to invent acceptable ways of monitoring the waste and, from a certain point of view, to take care of it. If it is no longer a question of *making it last* here, it is still a question of problematic time; a time we have to worry about and deal with. The work of Başak Saraç-Lesavre provides us with some very stimulating pointers here.[58] It underlines the fact that, far from intensifying clear-cut positions that can only set those who are 'pro' against those who are 'anti', a certain proportion of waste management reveals the forms of attachment at work in the uses of nuclear energy. This is because such technology generates a large number of residues with a capacity to persist well beyond the horizons of an average human life, and thus require extremely complex forms of care.

Bernadette Bensaude-Vincent has made a similar point:[59] she compares the different solutions used today in the management of radioactive waste, and shows that, from permanent or reversible burial to maintenance on the surface, each solution articulates different ways of 'dealing with' these materials, whose own temporality exceeds our own. But this 'dealing with' is not entirely passive: each solution shapes a different time based on the material and its propensity to last. Not all are equal: Bensaude-Vincent explicitly takes sides in her description. Probably disturbing some of her readers, she indicates her preference for projects that leave waste on the surface. These in fact have the great advantage of not putting off the question of maintenance until later (by focusing on the capacity of the soil to remain unchanged, or on scientific and technical progress), but of keeping it open and ensuring its relevance. Waste that is always visible near humans forces them to remain attentive. This waste demonstrates its *presence* in the double sense of the term: always there, it 'thickens' the present and prevents everyone from projecting themselves into too distant a future.

It is also this material force that the idea of stubbornness highlights, a force which sometimes owes less to the thing as a whole than to some of the materials that compose it and to their relationships with the matter of which humans are made, but also with everything moving through the environments in which they are found. The example of nuclear waste, and more generally of all the materials that haunt the world long after objects that had been presumed to be solid and inert have disintegrated, reminds us not only that the dance of maintenance brings together many partners, but also that it can take very different turns from what we first imagined when pondering the care of things. Stubbornness, a 'line of flight' of maintenance, helps us to recognize the subtleties of an art of 'dealing with', understood not on the level of the somewhat pallid fatalism to which the expression can sometimes refer, but on the contrary as the part of material entanglements through which the common becoming of humans and things is pieced together and brought to life.

From time to thing

The contrast produced by the ideas of prolongation, permanence, slowing down and stubbornness is not intended to support a structuralist analysis of maintenance. Other problematized relationships with time can obviously be identified in configurations very different from those described here.[60] If there is one take-away from this chapter that we would like to emphasize, it is the variety of situations that we have brought into dialogue with each other. The proposed itinerary serves above all as an invitation to sharpen our sensitivity to the many different ways of making time a problem, even when these are not the subject of explicitly stated principles. Each situation of maintenance can be observed in terms of the more or less sophisticated temporal *agencement* at work within it, by which things manage to last.

The configurations explored in this chapter form different 'timescapes'.[61] They form situated continuities which arrange the past and the future without ever escaping from the present, which they densify with many different activities and concerns. And as they make things last, they thereby make things *exist* in different ways. This is what we suggested at the beginning of this chapter with the story of the two clocks, highlighting the implications of the incredible affair of the clandestine restoration in the Panthéon. By suggesting that the members of Untergunther on the one hand, and Pascal Monnet (the new administrator) on the other, had perhaps not made quite the same clock last, we implied that, as well as the political and practical aspects of maintenance, there was an ontological dimension at play. Each situation described throughout this chapter has confirmed this intuition. Making something last comes down to defining in practice what matters in that thing, identifying what concretely needs to be taken care of to ensure that it is always the *same* thing that lasts. Can this or that part be changed? Should the colours be left to fade? How far should the functioning of this object be tested? How unchanged should the exterior shape remain? Each problematization of time and each form of maintenance organized around it actualizes the specific modes of existence of things. In many cases, these ontological gestures are self-evident. No one pays attention to them; they raise no disturbing doubts. Sometimes they become the focus of worry, as we saw in the previous chapter. Endless precautions are then deployed, and each intervention must take into consideration the unique features, those considered to be essential, of what is meant by 'to last'. At other times, the ontological aspect of maintenance becomes the subject of passionate debates, and even open struggles. Times with distinct horizons, and things with incompatible contours, then collide.

6

Tact

In *Experts et faussaires* (*Experts and Counterfeiters*), Christian Bessy and Francis Chateauraynaud focus on the fate reserved for a banknote during a fairly ordinary scene, at the checkout counter of a tobacco shop.[1] While one of them holds out the note in question, the retailer, who takes it in his hand and is about to cash it, pauses. 'With a dubious pout, he slowly raises his head, stares at us, and crumples the note for a moment, rolling it between his fingers; then, with the air of a connoisseur, he intones: "Gentlemen, this note is a fake, do you have another one?"'[2] The misadventures of the banknote highlight the practical implications of the trials on authenticity, which the authors show are never self-evident. They generate doubts and friction and require the deployment of different kinds of instrument in each case, as well as specific sensory skills.

In their study, Bessy and Chateauraynaud use the question of authenticity as a particularly effective way of highlighting the richness of the perceptual foundations of judgement. Drawing inspiration from their work, we would like to show that authenticity also offers a valuable means of raising awareness of certain very concrete aspects of maintenance. While it is not simple to rule on the authentic or inauthentic character of an object, what can we say about the anxieties that plague people who strive to ensure that the things they take care of *remain* authentic? What does this concern commit them to, in concrete terms? How do they manage to make the thing's authenticity last along with the thing?

In certain social sciences, the idea that people can be committed to the authenticity of certain objects and even seek to preserve it is ambiguous, not to say suspect. The notion of authenticity is itself problematic – one among those that students must learn to be wary of. Authenticity was once elevated into a safe haven, as an irreducible territory in the face of the triumph of commercialism, but authenticity in fact quickly became a commercial quality like any other.[3] Today there is a glut of works that denounce this 'usurpation' and reveal the fallacious nature of the numerous forms taken by the contemporary promotion of authenticity, whether in marketing, where certain people have become masters in the art of selling it, or in consumption. The critique is relentless. It blames marketers for being cynical, and consumers for being either naïve or even more cynical when they invest in an artificial authenticity so as to stand out from the crowd.[4]

It is obvious that an investigation into the care of things cannot be satisfied with this position, unless it condemns itself to systematically mocking the efforts of those who strive to make some objects last, sometimes even the most banal ones, in the interest of preserving what they consider to be their authenticity. Our interest in maintenance operations requires us to shun this suspicious outlook and to take seriously the concerns of those who care about certain distinctive features of the things they take care of. This is what allowed us in the previous chapter to identify important ontological discrepancies, including in the maintenance of things quite explicitly treated as authentic. Even more, by giving proper consideration to the question, we can leave aside the overarching explanatory vocabulary of the symbolic resource, or that of 'value' – whether in economic, heritage or sentimental terms – generally used in analyses of authenticity. We can stick to the kind of words used in all maintenance situations when confronted with this question, of the kind we have already encountered: words such as worry, concern and sometimes even anxiety. Because it is in this respect that the world of maintenance contrasts with the world of sale and consumption (in particular when the latter is reduced to the act of purchase, or even more abstractly, to the 'preferences' so dear to orthodox economists): to maintain is to be concerned. And when authenticity lies at the heart of the care given to things – which, let's not forget, is of course not always the case – it is far from being a value in which we could invest, or which we could simply benefit from by giving it a price. It's a problem. A host of problems, indeed – problems that haunt certain maintenance activities in diverse forms, often quite ordinary, sometimes trickier, and other times downright baffling.

We also know that the ontological gestures at work in maintenance are never carried out in some abstract realm. They involve direct contact with things, a more or less prolonged and more or less invasive contact that itself poses a problem: it is never only the definition of things 'as such' which is at stake in maintenance, but a certain kind of relationship between the thing that lasts and the men and women who make it last. A certain type of dance. In this dance, a question adds up to the twofold problem of time – the time that must be resisted and the time that must be brought into being – the question of the place that humans allow themselves to occupy in the life of things.

As we have noted, some forms of maintenance are modest, while others are downright pretentious. These terms helped us to outline a scale of the intensity of contacts between humans and things, and to ask whether the verb 'to make' in the expression 'to make things last' was really the right one. At the same time, we also showed that things themselves play varied roles in maintenance activities. By following these two lines of thought, we realized that the very existence of things in the configuration of their duration is staked on the always shifting, always fragile result of a distribution of action. But while we were able to get a first idea of the delicate nature of this distribution and its human, economic and technical cost, we have only scratched the surface of the question of the balance between these relationships. The maintenance of things that are considered authentic provides valuable ground for exploring this question further. First, it seems to push to its limits the problem of what the different roles are. How much should humans make their mark in order to keep things authentic? Does this authenticity not lie, on the contrary, in the autonomy of matter, freed from human intervention? Actually, if the care for authentic things is worth paying attention to, it is precisely because it is never reducible to these binary, disembodied dilemmas. On the contrary, a subtle art of hesitation is being cultivated in them – an art from which we have much to learn.

Adjustments

To grasp the concerns that weigh on the maintenance of authentic things and understand how they manifest themselves in different situations, we propose to return to the owners of old Mustangs as studied by Cornelia Hummel.[5] Authenticity is an essential dimension of the attachment that these women and men feel towards their cars and a key element of the care they lavish on them. It is not uncommon, for example, that over the course of their relationship with their automobiles, some people engage in restoration projects to retrieve missing or worn-out characteristics, such as paintwork, an element often considered as emblematic. But authenticity is also a source of daily torment for them. In particular, it increases the anxieties linked to breakdowns and even simple routine maintenance interventions. To the financial burden that the maintenance of such a vehicle represents for most owners is added a latent risk, which hangs like a shadow over each visit to the mechanic: that of having to carry out an operation – a simple replacement of parts or a more extensive repair – that would cause their car to lose its authentic character. Changing brake pads, replacing headlight bulbs or windscreen wiper blades: what are mundane interventions for most people often assume an existential turn here. The automobile entrusted to the good care of the specialist, or directly taken in hand by its owner (as happens in the world of old cars), actually risks, at the end of the operation, disappearing. Not that it will have completely disintegrated, of course, but the modifications undergone could irreparably cut it off from the world of Classic Mustangs. It will then be just an ordinary car.

In other words, by repairing it too much, by maintaining it too much, the owners risk destroying the very object of their care. Make no mistake, however: this disastrous outcome is quite hypothetical. If it is useful for understanding the particular tension that weighs on the owners of Classic Mustangs, this is not because it regularly happens to those who, although generally having modest budgets, devote a significant part of their resources to the maintenance of their car, but because it fuels a permanent concern. It is not so much the actual danger of the ontological break that matters here as the series of very concrete accommodations in which the concern for authenticity is expressed: the waltz of repeated hesitations which, far from representing a contextual and external dimension weighing on the ways owners take care of their vehicles, is the very heart of the matter, the principle at work. By reading the testimonies that Hummel presents in her book, we realize that, rather than just being a sword of Damocles hanging above each maintenance intervention, authenticity is a quality that is continuously cultivated. It is activated, and invented, in each of the adjustments to which these interventions give rise and, perhaps even more, in the very doubts which accompany them.

It must be said that the case of the Classic Mustang makes these doubts particularly noteworthy. If we had to stick to aesthetic aspects that could be appreciated from afar, and if the authenticity of the said cars could simply be admired in a garage or an exhibition hall – in short, if it were possible to turn them into museum pieces – things would undoubtedly be simpler. But the art object that is only good to be looked at is a turn-off for Mustang owners. Victor, one of the characters in Hummel's book, expresses the obviousness of this fact very simply: 'In automobiles, there's "mobile", so they need to move.'[6] Immobilizing the car to facilitate the maintenance of its integrity would be an absolutely unacceptable ontological break. The authenticity that must be maintained is inextricably linked to the multiple pleasures of regular driving, itself considered authentic.

It's now clear that we mustn't fall into caricature and draw a hasty contrast between the maintenance requirements of a running car and those of a work of art in a museum, as if the latter could preserve itself. It would be superficial, in particular, to decide that it's the mere fact of being 'used' that separates the former from the latter and makes their maintenance more difficult. In the museum, the works on display are also being used, simply by being exhibited. As we have seen, this use contributes to accentuating their fragility and presents numerous maintenance challenges. That said, it must still be recognized that these challenges remain relatively limited. The use in question allows for a fairly comfortable grip on the environment in which the works are kept – an environment that it is possible to control to a very significant level if the right means are devoted to it. The use of Classic Mustangs provides nothing of the sort – quite the contrary. It confronts the owners with very uncertain conditions, which sometimes present considerable risks. Obviously, they take precautions. They don't drive in winter, when ice and snow make the roads impassable for their cars. They're also wary of parking lots, as the people interviewed by Hummel tell us, and deploy many cunning tactics to ensure that nothing happens to their beloved car while they are shopping.[7] With regard to the concern for authenticity, driving thus represents a permanent danger, which explains the effervescence of attention described in chapter 3.

It would therefore be too simple to clearly separate the two requirements, as if the two words in the phrase 'a classic car' doomed owners to be torn between two clearly distinct normative horizons: one oriented towards preservation, the other towards use. Driving is not just one principle as opposed to another. Above all, it is not just a factor of extra fragility. It is also a vector of maintenance as such. Even more than most recent cars, running old cars is an integral part of their maintenance, while a prolonged period of immobility is synonymous with accelerated deterioration. After a few weeks, the liquids present in the vehicle (in particular oil and petrol) stagnate and form deposits of materials which can suddenly clump together in the engine when restarting, causing significant damage. To avoid this, in winter, owners take care to regularly 'wake up' their car which they've been keeping sheltered in the garage, in order to take it for a spin round the neighbourhood and reactivate the movement of the mixtures in the circuits. The same goes for the tyres, which don't cope well with the weight of the Mustang resting in exactly the same place for several months. The few winter outings are not enough to ensure that they do not inadvertently return to exactly the same position when they get back home. To avoid this unfortunate eventuality, some owners get into the habit of marking with a felt-tip pen the location of the point of contact between the tyres and the ground; others place them on special blocks; and yet others place the car on jack stands so that they don't touch the ground.

In short, Classic Mustangs have to be driven. They have to 'live', as almost all the owners interviewed by Hummel nicely put it. But how, in these conditions, can you ensure the authenticity of your car? How can you reconcile this life as a machine to be driven around with the concern for integrity? Isn't the task impossible? It is of course, if we consider authenticity as a monolith, a property with previously defined and immutable contours. However, the activity of Mustang owners demonstrates quite the opposite. These owners oscillate constantly, with many precautions, between an authenticity that is just suffered, experienced as a constraint with which one must comply, and an authenticity that is adjusted, flexible enough to withstand the accommodations that must be carried out if they are to continue driving a Classic Mustang.

Among these adjustments there are obviously security issues to be considered. If there is one area in which the automotive industry has been radically transformed in just a few decades, it's risk reduction. And in many ways, Classic Mustangs represent dangers on wheels, when compared to any modern car. Drum brakes are at the top of the list of safety concerns. As one of the owners humorously remarks: with this braking system, the car 'slows down', but you can't really say that it's braking.[8] While the sensations specific to these cars, which are free of the now ubiquitous additional electronic assistance, lie at the heart of the owners' passion, these owners rarely resist the need to install at least the brake discs that are now the norm. Changing the braking system obviously opens a breach in the integrity of the car but, all things considered, the modification is generally accepted. It doesn't stop this same owner who carried out the replacement asserting that everything in her or his car is 'original'. In addition to the brakes, seat belts are also subject to modifications: the simple 'belly' version present on many models being replaced by a 'three-point' version comparable to that found in more recent vehicles. Likewise, many owners change the headlights to improve visibility, as the discrepancy between the original lighting system and that of the vast majority of current vehicles is significant.

These accommodations, carried out in maintenance, or during repairs or restoration operations, are far from trivial. They are the subject of discussions, even debates, in which the doubts and hesitations of the owners are expressed. These deliberations take place with the mechanics, of course, but also within a community of amateurs, who exchange ideas online via a discussion forum and meet regularly at trade shows and gatherings open to the general public. Some of the adjustments are hotly debated. Others rely on more or less shared conventions. The most important of these is undoubtedly the convention that relates to options. In the 1960s, Mustangs were customizable cars. Buyers had a wide choice of options on each model, ranging from air conditioning to the installation of a rev counter on the dashboard, as well as the upholstery. This multitude of options is an important resource in the quest for authenticity of current owners, whose decisions are made on its basis. The optionality of the contemporary market thus draws around the notion of origin and around the idea of integrity – the main concern – a sort of expanded perimeter: a space in which choices can be made without too many uncomfortable contortions. If it turns out that, at the time a model was marketed, it could have been equipped, through an explicitly available option, with front disc brakes, the contemporary owner can decide that they are entitled to install this braking system without calling into question the authentic character of their car. It is by invoking this principle that one owner explains that she had power steering put on her car, because she was suffering from tendonitis so severe that it hampered her driving. 'Optional at the time', the function could be installed more than forty years later without anything becoming 'unnatural' or 'denatured', as she herself put it.[9]

As well as period options, a principle of reversibility comes into play. An intervention which conflicts quite blatantly with the car's integrity is in fact deemed acceptable as long as it is not irremediable. Hummel explains that an owner can change the rear light system to include a separate indicator, or even install a modern fuel filter, since it will be possible, if necessary, to reverse these modifications without too much difficulty.

Options and reversibility somehow relax the requirements of authenticity by expanding the contours of the physical integrity of the car that needs to be maintained. They offer owners room to manoeuvre within a set of constraints and uncertainties which are still significant. For certain cars, it remains very difficult to find original parts when they break or get too worn down, though. Opportunities are not easy to find, and stocks of new products, recovered from bankrupt factories or warehouses, are becoming increasingly rare. In some cases, more than availability strictly speaking, it is the price of these pieces that prevents owners from completely sticking to their pet ideal of integrity. If for many 'consumables' such as filters, gaskets and spark plugs everyone agrees that modern parts can be used, in other cases there is more hesitation. The engine parts, the steering wheel, the bodywork elements... that there is no consensus on these components is self-evident. All of them confront drivers with their own involvement in the becoming of their car and its authenticity. By taking care of their Classic Mustang, they transform it – especially since the world in which cars and their owners operate is changing too.

Indeed, in addition to road safety aspects, environmental issues are upsetting the trade-offs that were until now possible to make to ensure that Classic Mustangs can circulate on the roads of Europe without losing their authenticity. In France, for example, many cities already ban the most polluting vehicles. If the registration document specific to classic cars authorizes them to be driven while exempting them from numerous legal obligations (such as the presence of a safety belt that complies with standards), this will not be the case for regulations relating to the environment. When they find themselves in a position to decide, generally following a major breakdown that requires the engine to be changed, owners are faced with a dilemma. Should they anticipate a strengthening of regulations? A generalization of bans? Will pumps distributing petrol without the addition of ethanol become so rare that they will only allow a limited number of trips to be made? According to some owners, there is little doubt that, in the coming years, there will no longer be any question of a dilemma, and a modified engine, running on unleaded petrol, will be one of the components of cars that will still remain authentic Classic Mustangs.

So we are a long way from the cynicism and distrust that authenticity inspires in some people. The care taken with things whose authenticity people strive to preserve absolutely shows how delicate the matter is. It also underlines the ambiguity of the position of those who lavish such care. While they of course take pleasure in being around these things and using them, they are no less constrained by them, never completely free to do with them as they please. In her book, Hummel also borrows the term 'attachment' from Antoine Hennion to describe the relationship between Classic Mustangs and their owners. In line with the pragmatist philosophy of William James,[10] talking about attachment is a way of understanding both the material and the personal part of the relationship to things. This rich and dynamic relationship cannot be reduced to the symmetrical dualisms of technical determinism on the one hand (where everything is conditioned by the force of technology) and interpretive flexibility on the other (where all value is symbolic and conventional, produced by mental and cultural 'representations' alone). If the notion of attachment was useful to illustrate how the scientists of the space agency had become 'obliged' to the spacecraft which continued to head out into space years after the scheduled end of their activities, we can see that it is also valuable for understanding the particular aspect of the care of things highlighted by the Mustang owners' concern for authenticity. The term holds together the apparent contradictions of the experience of care, refusing to choose a priori between pleasure and constraints, between human will and material resistance. Because it prevents us from deciding between causes and intentions, between action and passion, the term 'attachment', simply put, offers the possibility of taking seriously the ambiguities of people's relationship to things without emptying them of their richness. To maintain is to simultaneously attach the thing we are taking care of to ourselves, and to attach ourselves to it.

Insisting on the constraints involved in the relationship that develops between Classic Mustangs and their owners should not, however, lead us to believe that the authenticity at work here is specific to the domain of 'amateur' practices alone. Even if, as we have seen, it is possible to condition certain works in such a way as to slow down their ageing as much as possible, doubts, hesitations and the adjustments that have to be invented in situ are also omnipresent in the field of artistic conservation and preservation. Ever since Duchamp's ready-mades, contemporary art has even created ever more situations of uncertainty and sources of questioning.[11] Installations whose boundaries are difficult to demarcate, hybrid machinic devices, the use of living materials: works of art confront curators, who one might think are attached to forms of authenticity much less flexible than those of Mustang owners, with real ontological puzzles. How can we maintain the objects 'without qualities', drawn from the world of public commerce, that make up certain works? Can we replace this or that mechanical element, now unavailable, with a more recent one without forever affecting the authenticity of an installation? And how can we preserve digital works, whose writing and reading interfaces so quickly become obsolete?

Here, the requirements of authenticity are obviously not in all respects identical to those which concern Mustang owners, if only because in large Western museums a tripartite relationship is generally established between the works of art, those who take care of them, and those who create them, the artists. That said, the dynamics of attachment and obligation are very similar. Maintaining authentic things over time means confronting doubts, cultivating hesitations, deliberating and, more than anything, composing with the materials. And, as we have discovered throughout the previous chapters, there is nothing inert about these materials. Composing with a material does not simply mean that one can simply focus on what its physical properties (its greater or lesser fragility, in particular) allow or do not allow one to do. This is particularly clear in the practices of Classic Mustang owners. The car is never a materially stable 'given' in the relationship. What owners experience of its material characteristics, what they know about it, continues to evolve over the course of maintenance interventions.

Surprises

So far, we have highlighted the quandaries of Mustang owners and the adjustments they are prepared to make. We have suggested that these adjustments consisted of negotiating acceptable discrepancies between the operations that it was physically possible to carry out and an ideal version of the car, perfectly identical to the one leaving the Ford factory the year it was placed on the market. However, on this front too, things are actually a little more complicated. Hummel's long-term investigation shows that it is in fact quite rare for owners to have a clear and definitive vision of the initial state of their car, a vision which could serve as a frame of reference for evaluating the extent of adjustments that can be made. Purchased second-hand, and generally imported from the United States, Mustangs have often already lived several lives. Most owners have no way of knowing the details. Equipped with a few official documents – for the lucky ones at least – many people have only a fairly vague idea of the origin of the car they are purchasing. The question of integrity, so important in the ongoing development of authenticity, is therefore itself fragile. It is above all subject to regular revisions which stem from the other side of maintenance activities carried out under constraints of authenticity. Indeed, if interventions related to the maintenance and restoration of the car always represent risks, they also offer opportunities for sometimes unexpected discoveries.

We had the good fortune to talk for many hours with Cornelia Hummel, who shared with us fascinating stories not included in her book, but which we analysed together in an article, some details of which we include here.[12] This led us to discover the tribulations of Margaux, the owner of a 1964 1/2 Mustang, who makes a brief appearance in Hummel's book. When she decided to re-paint her car, Margaux followed the operations closely. The sanding, she says, was a delicious archaeological process, with the car going through the brand's famous official colours: Silver Blue, first, then Caspian Blue, Silver Blue again and finally a rather unconvincing red. Little by little, the Mustang revealed its successive changes, reflecting the decisions or indecisions of its previous owners. But a surprise awaited Margaux and her mechanic. While dismantling certain parts for the operation, they found, stuck between the windscreen and the sheet metal, pieces of fabric of uncertain origin. Once the sheet metal had been completely sanded down, brown spots appeared that were not rust. Upon further information, the mystery was partially resolved: the car had undoubtedly had a vinyl roof when it left the factory. Until then, the owner had been quite categorical about the origin of her car. She thought she had enough information to know exactly what it looked like 'originally' and thus guidelines for her progressive quest for restoration. At the end of the sanding/disassembly operation, everything had changed. The car was no longer the same. Once the traces of another story were revealed and a part of its identity – hitherto unsuspected, and quite well hidden – revealed, it appeared in a new light. And this was not without consequences, of course, for the rest of the operations. Margaux also found herself in a dilemma. Not because an accident, wear and tear, or the desire to improve her driving conditions was forcing her to make difficult choices, but because the car itself had pushed her into a corner by providing material clues from its past life. To install or not to install a new vinyl roof? The answer didn't take long: there was no question of doing so. The disadvantages of using a new element (Was it watertight? Might it rust?) were even more significant than the constraint of a redefined integrity. The car would do fine without it. Hummel mentions another incident experienced by Margaux, who, while taking the 'Mustang' side badges on her car apart, discovered that they were not exactly the same length. On the discussion forum, one of the great specialists in the history of the brand told her the reasons: the two insignia probably did not date from the same year; the shorter one was undoubtedly the original one (installed in 1964), while the longer dated back only to 1965. The car had revealed another part of its history, to which Margaux decided, this time, to remain faithful while leaving the two slightly mismatched badges in place.

Sometimes the discoveries are even more staggering. Everyone knows, in the French-speaking community of Mustang owners, the story of the young Frenchman who inherited one of his grandfather's cars. The magnificent 1967 Wimbledon White coupe hadn't been driven for a long time and had to go to the mechanic before its owner could fully enjoy it. This is where the trouble began. Upon overhauling it, the mechanic discovered that the engine was that of an ordinary Mustang, but that it had traces of preparation for 'racing'. Although coming from a fairly standard model for the time, this car seemed to have been involved in car races for part of its life. Having gone in search of explanations from specialists and historians of the brand and motor racing, the mechanic returned with a surprising revelation: the car had belonged to Johnny Halliday – one of the greatest celebrities of French rock 'n' roll – who had raced under the colours of the Ford France team in 1967. Once the discovery was confirmed, the mechanic's intervention completely changed in nature. It was no longer a question of getting one Classic Mustang among others back into working order, but of taking care of Johnny's car. Above all, the question of integrity became much thornier, as the car had somehow lived two, even three, lives. The owner and the mechanic finally decided to restore this legendary Mustang to the appearance and engine it had when it belonged to the singer, going so far as to share the costs of a restoration well beyond the owner's initial budget.

This dynamic of discoveries is very important if we want to understand the doubts and hesitations that can accompany certain forms of maintenance, including in situations that are a priori the least equivocal, where it's 'sufficient' to maintain the authenticity of the thing we are taking care of. It complements what we have learned about the forms of material exploration in which maintainers engage. It also adds a new sheen to the resistance and indeed the obstinate desire of things to last, as we described them in the previous chapter. Here, actually, things resist less than they reveal themselves. They themselves play a part, sometimes via tiny traces, in the authenticity that maintenance both preserves and fashions. Things remind us that they too have their say.

In an article dedicated to the restoration of part of the old University of Vienna, Albena Yaneva focused on these 'surprises', where things spring up for those who take care of them.[13] At the beginning of the 2000s, she studied the vast renovation project of this seventeenth-century building, entrusted to the architect Rudolf Prohazka. In this type of project, many specialists plan the operations, armed with countless archives that help them – much more than the modest owners of American cars of little value – to identify the original elements that the operation needs to restore and protect. However, these documentary elements are not enough to completely programme the intervention, step by step. A building always opposes its own renovation in one way or another. It is a 'disobedient' object, which sometimes resists. But these resistances are not all negative and cannot be reduced to the physical difficulties that maintainers may encounter in interactions with things as we described them previously. They can even be very productive and give rise to epiphanies. During its restoration, a building sometimes reveals to the many people who are busy inside and around it elements which they did not know existed. By 'resisting', it then adds new pages to the historical documents and archaeological studies, with details drawn directly from its walls and ceilings. Sometimes these surprises are welcomed by renovators. Yaneva mentions in particular a wall which seemed on paper to be 'of historical value', but which hampered the architectural project and turned out to be much more recent than imagined, ultimately becoming a candidate for destruction. Other surprises tend to disconcert restorers: one period material is less refined than had been thought; some structural elements are very flimsy and definitely do not conform to contemporary requirements... Finally, others, like signs of modifications to the car's engine which lead to the discovery of the identity of its prestigious previous owner, can produce explosive revelations. In the case of the Vienna University building, the biggest surprises arose on the second floor in a room whose painted ceiling was at the forefront of everyone's attention. One morning, during her regular visit to the building, Yaneva found the room empty and the construction site silent. As she approached, she realized that the walls had been scratched and that the few areas thus cleared had revealed the elements of a fresco that nobody knew existed. The building had spoken. And what it had revealed was so unexpected, so important in the quest for authenticity that characterized the entire project, that it was felt necessary to suspend the interventions and take the time to revise the plans.

This is one of the important dimensions of the surprises that things can have for us when we take care of them. They are not simply 'informative'. They not only allow people to complete the image they have of them, of their history and their origin. Sometimes surprises interrupt the care. They throw maintainers off track. They confront them with new choices, forcing them to make decisions that were unthought of before their emergence. When they are sufficiently disturbing, these surprises even force those who experience them to carry out much more in-depth investigations than those they had previously conducted. The discovery of the 'rebel' mural, as Yaneva calls it, thus inaugurated a new series of investigations carried out by new experts. These gradually solved the mystery of the paintings, which had probably served to frame the stage on which public experiments were conducted in the seventeenth century, before being later hidden by curtains when the same stage was used for the performance of plays.

Just for once, let's not linger on the details: what matters here is the opportunity for an inquiry triggered by the material surprise. This calls for a profound reconfiguration of a territory of authenticity that had seemed to be stable. As suggested by the case of Johnny Halliday's coupe or the research undertaken by Margaux and her car mechanic, we also find investigations of this type among Mustang owners, many of whom go in search of new elements of the history of their car after discovering a clue leading them to suspect that they still have a lot to learn about the episodes of its life, or even its origin. The dynamics of the various emergences and inquiries involved show again that, in the scenario that interests us here, each maintenance intervention is indeed a trial of integrity whose outcome may entail adjusting the contours of authenticity. The latter is not available in the form of a pre-established plan, a series of criteria or general principles that simply need to be applied. The authentic thing cannot be mechanically 'preserved' or 'conserved'. It is always *to be made*, in the beautiful expression that Étienne Souriau uses about works of art.[14] But the surprises also show that this form of generative action is not the sole responsibility of maintainers. Maintenance in terms of authenticity is a matter of transactions, of successive reactions, sometimes from the thing, sometimes from those who take care of it.

So here we are again, facing the dance of maintenance and the shared movement that it accompanies, gesture by gesture. However, sensible encounters here appear to be caught up in a very particular tension: that generated by the necessary adjustment of gestures between the different types of dancers. What balance should we find in this relationship? What proportion should be left to the movements carried out by each partner? In the previous chapter, we saw that each way of making time a problem corresponded not only to a certain ontology of the thing to

THE CARE OF THINGS

be maintained, but also to the more or less pretentious position of the maintainers. When linked to the metaphor of dance, the expression amounts to saying that certain dance partners tend to want to do more than others – that they seek to 'lead', sometimes whatever the cost. When one is concerned with authenticity, however, this attitude has its limits. As Yaneva points out at the end of her Austrian investigation, renovation, however ambitious, cannot be accomplished in the form of a 'heroic battle' that humans wage against matter. We must accept that the thing will also take a few steps, or even sometimes lead the movement, otherwise it will simply jump track. We must even recognize that the thing we are taking care of sometimes acts as a 'moral agent on its own'.[15]

This matter of balancing positions and adjusting the intensity of each person's action is far from simple. However, it is crucial. In our exploration of what maintenance can teach us, it represents an important step since it enriches the ontological question (with what mode of existence does a form of maintenance associate the thing maintained?) with a relational dimension (to what and with whom is the thing associated if it is to exist in this way?). It also lies at the heart of numerous debates within professional worlds where the issues of authenticity are fundamental. It is towards these that we now propose to turn.

Heritage diplomacies

In her very fine documentary on the last phase of the restoration of Chartres Cathedral (between 2014 and 2016), Anne Savalli allows her camera to linger on a few key sites of this vast operation.[16] Inside and around the cathedral itself, she observes those who take care of the sculptures, the walls and their decorations, the keystones, the organ and, of course, the stained-glass windows. She also ventures into anonymous rooms, where meetings are held between the various site experts and the sponsors, and spends a long time in the workshops where the stained-glass windows are transported to be cleaned and sometimes repaired, before being put back in place. We discover the diversity of the gestures, positions and tools, and of the words and professional expressions that nourish this labour of collective care, made up of a multitude of meticulous interventions, most of which seem tiny compared to the monumentality of the place.

As privileged witnesses, we see the first layer of dust disappear from a column, swept away by the energetic brushstrokes of a woman restorer wearing a protective suit and an impressive mask. We are struck by the appearance of a few centimetres of the layer of thirteenth-century paint, now resurfacing 'in perfect condition', while another restorer scrapes the wall with an instrument barely larger than a scalpel to remove what she presents as a whitewash from the nineteenth century, then a layer of paint from the fourteenth century. She then points out some marks discovered during the previous weeks: inscriptions traced by the hand of the first masons, which everyone had agreed to preserve.

In the workshop, an item of stained glass, whose careful removal we have followed, is observed, handled and brushed before being evaluated by the person responsible for the operation. Equipped with a notebook, using her eyes and fingers, the expert identifies each piece of painted glass by assigning it a period, revealing to our lay eyes a heterogeneous historical mosaic which will be the subject of differentiated treatments. We then hear her, with the restorers on her team, counting the 'breakage leads', the connecting elements that were not present in the original design of the stained-glass window but were added following a breakage. Later, we witness how one of these leads, which had cut through the depiction of a face, is replaced by a line of transparent silicone; the restorer shows, by moving the piece of glass, that this has the advantage of maintaining a certain flexibility, and is, even better, almost invisible. Freed from its big black scar, the face is transformed.

Back in the cathedral, we are amazed by the action of the layer of latex previously applied to the statues, which, once removed, takes with it most of the dirt that has built up right in the folds of the stone, revealing a dazzling whiteness without any corrosive action being carried out. This technique is the most 'respectful' because it treats only the surface of the sculpture, as the people who use it explain. We also follow the movements of a restorer who works, in a very uncomfortable position, to reveal the paintings appearing under the layers of dust and dirt at the top of the arches. She seeks to reconstruct the original decorations – a stratigraphic operation that is far from simple, as what appears (colours and drawings) seems tiny.

All of these scenes are wonderful, and Savalli's film provides a perfect complement to the various points discussed so far. We come across fragile materials, and attentive, sometimes hesitant, human beings who work on the temporality of the monument using a flow of heterogeneous materials, leaving room for the unexpected to emerge. Those men and women, busying themselves in the smallest spaces of the building, take part – with their repetitive, controlled and concentrated gestures – in a great choreography composed of numerous steps regulated in advance, and varied improvisations that are even more numerous. Finally, since this documentary dwells on the subtlety of the interventions, and also shows the sometimes concentrated and sometimes loving way the restorers gaze at the different elements of the cathedral they are taking care of, it is a precious illustration of the very idea of 'taking care of things'. But if it is so fascinating, it is also and above all because it does not stop there. It addresses an aspect of the restoration programme that is almost invisible in daily material operations, although it is obvious as soon as one pays attention to the discussions between experts. Two of these conversations appear in the film. They open a breach in the almost idyllic landscape of coordinated interventions carried out by a multitude of industrious women and men, participating in the same vast operation of care.

The first scene is somewhat ambiguous. People in everyday clothes, whose professions or status we do not know, are walking along the bays and seem to be debating in relatively hushed terms. We only witness a short extract of the conversation, which is enough to alert us. It is about the decorations, which, as we realized earlier, were not easy to reconstruct. One woman, who seems responsible for part of the project, explains: 'The problem is that if we restore the decorations there' (she points to an elevated area) – 'and we know that we can do so, it will be very different there' (she points to another area). 'And at the transept, we have nothing left, just a few tiny bits. And in the choir, on the other hand, we have the gilding and everything. So' (returning to the first zone), 'we still have to see it within the space of the cathedral. There's the archaeological aspect, and then the architectural and volumetric aspect. So it's complicated.'

A few minutes later, a man comes up with a proposal which may be deliberately provocative or not, but seems to make everyone uncomfortable: 'Why not commission a new decoration from an artist? There was some desire to mark the keystones – we can try to keep the principle, and have this axis leading towards the choir' (he gestures with both arms to accompany his explanation). 'And as we're thirty metres away, we can allow ourselves a certain imagination, the detail won't be seen. If a colour is a centimetre or so out, it won't matter.'

After a silence, someone retorts: 'Yes, but we have no idea of the colours here.' The man replies: 'Exactly. That's why we distort our perception of the building less than if we put nothing in at all, in my opinion.' The scene is interrupted by an exchange of embarrassed glances among those present.

This contrasts significantly not only with the previous restoration scenes, but more generally with everything we have highlighted so far. Even when we identified distinct forms of maintenance by pointing to specific ways of making time a problem, we were in fact mainly concerned with describing care activities that were aligned with each other, forms of collaboration from which neither debate nor any profound disagreement emerged. The doubts and hesitations that we focused on when looking at the maintenance and restoration practices of Mustang owners did not reveal the kind of discrepancies that this curious conversation suggests. Quite the opposite: if Hummel mentions in her book a few debates that liven up the discussion forum, she especially emphasizes the benevolence of the exchanges and the capacity of the community, online as well as during trade shows, to produce consensus in the collective establishment of authenticity. This first, somewhat strange, discussion leads, in Savalli's documentary, to a different line of argument that becomes clearer in a second scene during which other protagonists make much more explicit contributions.

We are once again in the bays of the cathedral. Several women are chatting; all of them, we guess, are experts in different areas of heritage conservation. The first reports on the 'hypothetical reconstruction' which was carried out on the polychromes of the keystones, which are, she says, all in a 'very incomplete' state: 'There probably were decorations, but we can't see them. There's a second polychromy, with gilding. We can see some flowers. On the former we can recognize background colours, but we don't know if these colours have been restored or not.' Another woman intervenes: 'I'm sorry to barge in, but these restorations, these reconstitutions of gilding that look a bit flashy... well, it's a very long way from the theories of conservation as they're usually thought of, and it makes many archaeologists cringe, particularly in other countries. Here we still have a case where, for the two decorations that are still more or less present, the question of the conservation of both of them arises – might it not be an opportunity to say about these bays that we can try something else, and we don't restore them, or less, so as to conserve the keystones more?'

Her terms are more precise than those of the previous scene. They clearly show that the entire restoration, which until now has mainly been followed in contact with the materials, involves longstanding problems, and is the object of external and even foreign scrutiny; also, it goes against certain 'theories' that are presented as crucial to the profession nowadays.

A third woman speaks, without masking her annoyance. She vehemently explains that, if there has been any debate, it's closed now. The arguments she expresses are quite different from those put forward by the man who thought that an entirely new decoration could have been ordered; it's clear, however, that they're not completely foreign to the position he defended:

> We assessed this issue of conservation on several occasions. We have to realize that we're up at the top, but when we de-scaffold, we'll be at the bottom. When we deal with conservation, with just the archaeological traces, which is what we do in many monuments, we actually have a completely illegible reading from below, completely incomprehensible traces that which sometimes harm the architectural elevation. Which means that instead of highlighting what's been transmitted to us, through our discoveries, there are times when they're just spots of archaeological colours... it has no unity. [...] And also, we weighed the possibility because we're in a building that's already been treated on the 'royal portal' part and on the 'choir' part, which means that we can't just embark in the middle and say to ourselves, 'We're changing everything, and simply treating elements of archaeological painting as matters for conservation', which, from below, will appear as an unfinished project, an incomprehensible project, and a factual detail that ultimately contributes nothing to the readability of the architecture. But here, if we just keep the traces of red with the gilding in the state of total oxidation where it is..., then, seen from below, we'll be asked why the project ended due to lack of funds, that much is obvious.

This tirade breaks the impression of harmony and tranquillity that had hitherto emanated from the myriad of maintainers present on the site. It immerses us in the much more conflicting universe of principles and frames of reference. The concern for authenticity takes a completely different turn here, and doubts and dilemmas crystallize in the form of in-depth debates pitting experts with clear-cut positions against each other. In the space of a few minutes, the scene bears witness to the main questions that fuel the force field of heritage conservation, an area that for decades has been riddled with controversies and existential crises. For example, we can observe in the terms being used an attachment to specific values, some of which are explicitly contradictory (the provision of factual knowledge, architectural readability, etc.). This resonates directly with the work of Aloïs Riegl,[17] who greatly influenced the whole field by providing an overview of the different reasons why certain monuments could be the subject of a conservation programme. But, at the same time, it is clear that these values do not explain everything – far from it, since it is also a question of what is actually available in this precise situation: what the cathedral has allowed to emerge through the stratigraphic operations, what has resisted humidity, exposure to light and, of course, the interventions of cleaners and other restorers who have succeeded one another over the centuries. The short exchange between the three protagonists shows that it is indeed a conservation policy that is at work here, one based on choices, organized as a strategy and thus opposed to other possible paths.

The whole library of the École des Mines, where we both work, would probably not be enough to house the books and journals that have fuelled the debates and disagreements about artistic and architectural conservation since the nineteenth century. This is a huge and controversial area, and it would be absurd, under the pretext of following the path outlined by Savalli, to claim to circumscribe it in the few pages that follow. However, it seems to us to deserve some attention – not out of any taste for the study of socio-technical controversies, nor for the pleasure of bravura rhetorical flights of fancy and definitive declamations, but because we consider that something other than simple positional games is manifested in these divergences and these controversies. In the debates, from the most technical to the most theoretical, the lines of tension which characterize the concern for authenticity can be seen. With them, differences in the concrete ways of organizing the care of things are expressed, and fundamental ontological questions are formulated.

It is obviously impossible to situate precisely in history, or even in prehistory, the emergence among humans of a concern for preserving objects of the past. As far as monuments and sites charged with collective memory go, we know that concerns go back at least to Rome,[18] and that the idea of a collective heritage to be safeguarded has been consolidated over the centuries, in various forms.[19] Most specialists, however, agree that in Europe, the question of heritage conservation underwent significant developments at the beginning of the nineteenth century.[20] A few decades after the French Revolution, a concern for the preservation of certain monuments was structured and institutionalized – monuments that were then given the official term 'historic'. In France, the High Commission for Historic Monuments was created by Mérimée in 1837. For the restoration of the Madeleine Church in Vézelay in 1840, he called on Eugène Viollet-le-Duc, still very young at the time, who would become one of the leading lights of the discipline. Over the century, the movement became more widespread. Many people saw it as the expression of a strengthening nationalist sentiment, combined with a growing desire for public education, and then, at the end of the century, the result of anxieties aroused by successive technical revolutions which raised fears of a collapse of society.[21] In just a few years, several countries created institutions and promulgated laws, making the protection of ever more numerous ancient buildings a matter of public concern. The twentieth century witnessed a meteoric acceleration of this trend, with a considerable expansion of the list of entities targeted by preservationist concerns and an expansion of the geographical areas affected. Doctrines, conferences and international charters followed in increasing numbers, and a world of expertise and specialized professions dedicated to making a myriad of things last by virtue of their memorial properties gradually took shape.

Over the course of its institutionalization, and then its spread, heritage conservation was shaken by scandals and controversies. From the city of Carcassonne to the Sistine Chapel, including the frescoes of the Basilica of Saint Francis of Assisi, innumerable projects have sparked controversy; in addition, there have been many disputes in the field of artistic conservation. Most of these sometimes heated debates are organized around a major split, one that can be traced in the discussion filmed by Savalli in Chartres. This split forms around what has gradually become established as two major ways of cultivating authenticity – two ways of practising the care of things that are now generally distinguished, at the risk of being a little simplistic, as either *restoration* or *conservation*.

On one side, we would find Viollet-le-Duc, the greatest (and one of the first) restorers. The principles to which he was attached are well known. He viewed restoration as an opportunity to give back a building its initial appearance. To this end, he advocated a rigorous, almost archaeological approach, and drew on a huge amount of documentation. From a certain point of view, Viollet-le-Duc completely accepted the need to make his restoration projects 'history workshops'.[22] In his view, scrupulously reconstructing the original state of a monument, itself understood as a document, consisted in behaving like a historian.

At the opposite end of the spectrum, we find two English authors: John Ruskin and William Morris. Both strove to defend a radically different version of the maintenance of ancient monuments by openly criticizing restorative interventions of the kind of Viollet-le-Duc, which they considered to be tantamount to an act of destruction. According to them, the only adequate way to make buildings and works last was to practise continuous and delicate maintenance. From this point of view, restoration operations were merely a symptom of neglect, the result of too long a period during which the monuments in question had remained without care, almost abandoned. The position of the two authors directly echoes the idea of slowing down that we described in the previous chapter. There was no question here of imagining that one could produce history by subjecting monuments to radical treatments. In Ruskin and Morris's perspective, '[o]ld edifices like living beings called for daily care, not artificial rejuvenation'.[23] Furthermore, the inevitable prospect of ruin and decrepitude looms over the horizon of such a view.[24]

What can we learn from this conflicting dynamic, which seems to have been at the heart of heritage conservation practices ever since their modern origins? It gave rise, with its successive variations, to several readings. One of them was very early based on the question of historical truth and accusations of facticity. Viollet-le-Duc and his disciples were criticized for producing a distorted image of history. Jean-Michel

Leniaud notes that Henry James, among others, wrote in his *A Little Tour in France* about the city of Carcassonne, restored at great expense under the supervision of Viollet-le-Duc:

> For myself, I have no hesitation; I prefer in every case the ruined, however ruined, to the reconstructed, however splendid. What is left is more precious than what is added: the one is history, the other is fiction; and I like the former the better of the two, – it is so much more romantic. One is positive, so far as it goes; the other fills up the void with things more dead than the void itself, in as much as they have never had life.[25]

The criticism is very familiar, and it still surfaces in the scenes of Anne Savalli's documentary: in the name of restoring a part of the truth, some people allow themselves to add elements that are currently absent. In the name of another version of the truth, they are accused of fabrication, counterfeit and lies. We can note in passing that one way out of this impasse was proposed quite early on, attributed to Camillo Boito, an Italian architect, who formulated it at the end of the nineteenth century in *Conservare o restaurare* (*Conserve or Restore*).[26] The method, which is still one of the main doctrines in this area today, consists of making conspicuously visible all the additions made during a restoration operation: this is the only way that no 'deception' will inveigle itself into the parts of the buildings whose youth is being allegedly restored. Not without provocation, Boito even explains that he prefers what he calls a 'failed' restoration to a 'successful' restoration, as the first, unlike the second, makes it easy to identify the additions.[27]

Another reading of the cracks in heritage expertise and practice focuses on the relationship with the past. This is a huge question and part of the criticism actually applies to the entire profession. The heritage frenzy that has gripped the world since the middle of the twentieth century is, in the view of some people, the symptom of an unhealthy relationship with time. Françoise Choay speaks of the 'Noah complex' to describe the tendency to make everything into a matter of heritage.[28] It is also this obsessive attitude to heritage that François Hartog stigmatizes with the notion of 'presentism'.[29] At a more specific level, since the first controversies, the problem of the past has regularly been brought up by the major theorists in the field. The question that drives most people is ultimately quite simple and can be expressed in these terms: how far back should we go? 'To the origins', would have replied Viollet-le-Duc, who initiated a practice that continues to be debated today: 'scraping', so as to rid the building being restored of elements that were not present at its creation and with which, over the centuries, it has adorned itself. This is precisely what we can witness in Savalli's documentary, when a restorer gradually reveals a thirteenth-century painting by removing more recent layers of whitewash from the wall she is treating. 'Why deny all heritage value to the time that elapsed between the origin and the moment of the restoration?', other people retort. This is the position, as we have seen, of Ruskin and Morris. It is also the one that gradually imposed itself as the profession became more institutionalized and international. From Boito's work, the idea was established that the successive transformations of a building, the interventions to which it had been subjected, could be considered as historical elements that restoration should not erase under the pretext of focusing entirely on the moment of origin. In the case of Chartres Cathedral, we learn at the end of the documentary, during what looks like the final handover of the works, that the restorers were surprised to discover in the bays two slogans from the time of the French Revolution, one displaying the word 'Republic' and the other 'Constitution for' (the rest hasn't been found). The site manager explains that it seemed important to them to preserve their trace, even though they were obviously not present at the origin of the building.

It is therefore ultimately the principle of granting a certain 'historical depth' to the preserved monuments that won the day, although the interest of the restorationist trend inherited from Viollet-le-Duc was also acknowledged. This is evidenced by many of the operations observed in Chartres Cathedral: most people agree that the building has 'rediscovered' the light and brilliance of the colours which bathed its interior in the thirteenth century and that no one had since experienced until this restoration. This historical depth has become all the more dominant in heritage practices as they have been equipped with ever more precise scientific methods taken from archaeology: these now complement the historical expertise of architects... and sometimes also add to the epistemic tensions in the field.

Issues of historical truth and the relationship to the past are obviously fundamental to understanding how very specific forms of care are discussed and modified in the field of heritage conservation. They bring with them a third question, one that often remains hidden below the surface of the most familiar debates, but one that is essential: the place of humans in the conservation process. To what extent do men and women have the right to interfere in the lives of the things they care for? What should their place be in the becoming of the material part of their own history? Here we encounter questions that directly echo the ideas of modesty and pretension that we discussed in the previous chapter. They played a crucial role in the evolution of concerns for authenticity that drive heritage conservation, and it is on them that a principle now omnipresent in the field was based.

Let's return to those contrasting views. Unsurprisingly, Viollet-le-Duc is the champion of human commitment. The historian's gesture that he defends is one of omnipotence, an authoritarian response which, armed with knowledge, can go so far as to impose profound modifications on the building as it presents itself. A famous quotation taken from his dictionary of architecture perfectly sums up how far this idea can lead: 'To restore a building is not to maintain it, repair it or remake it; it is to re-establish it in a complete state that may never have existed at any given time.'[30] Armed with the knowledge of the history for which he vouches, the restorer is entitled from this point of view to 'correct' the buildings. A restorer can not only rid them of features that have been added to them over the years, reconstructing those that have disappeared for one reason or another; they can also remove original elements considered as construction errors, deviations from the architectural spirit of the time. It is almost an anti-materialism that is at work here, a 'historicist rationalism'[31] in the name of which the idiosyncrasies of each building do not have to be taken into account, any more than do the geographical particularities that a national reading of the history of architecture, however erudite it may be, largely ignores.

There is much to be said about this way of seeing things and its implications, both epistemic and political. It is particularly important to recall the role that heritage conservation played in terms of national narrative in nineteenth-century France, when Mérimée and Viollet-le-Duc were at their most influential. But let's stay with the question of the place of men (and women, though they were under-represented at the time) in this process. Very early on, the model of the all-powerful restorer was the subject of debate which went so far as to call into question the professional qualities of those responsible for restoration. In his work, which describes the violence of the disputes of that time, Jean-Michel Leniaud discusses the positions of Baron de Geymüller. During his speech at the Third International Congress of Architects in 1897, this fervent opponent of Viollet-le-Duc linked what he considered to be the deviations of restoration to the idea of the creative architect: 'The conservation of historic monuments [...] asks them [the architects] to do violence to themselves – it demands that they, who are eminently progressive, if not radical, become momentarily conservative, eminently disinterested.'[32] In the summer of 1900, at the Fifth Congress, reminding listeners of his position, he added that it was imperative to train architects in conservation in order to help them get rid of their constructive impulses. 'We would accustom [...] all architects [...] not to have an exaggerated fear of every crack and not to agree too easily to demolish and rebuild, but rather to conserve.' Restorative architects must be educated in different forms of action. For conservation to matter, it must be detached from the framework of creation. The arguments are reminiscent of the terms of the 'maintenance art' of Mierle Laderman Ukeles, on which we have been drawing since the beginning of our exploration.

As we have seen, Ruskin and Morris, for their part, defend a radically modest position. While humans are connected with the destiny of things, this in no way gives them the right to impose their mark on them for the sake of an unconditional attachment to the past. Men and women, including the biggest names in the discipline of history, cannot claim to go back in time, imposing major material transformations as they go. As the providers of essential but humble care, they are at the service of objects that never stop living and thus ageing. Their place is in the background, and their actions need to be moderate. This position, intimately associated with the recognition of the historical density of ancient buildings, is quite close to what has gradually become the norm among researchers, experts and practitioners, as we find in the main doctrinal texts of heritage conservation. It has been crystallized in the expression 'minimal intervention', which today functions almost like a mantra in the field.

It is to Cesare Brandi, a central figure in worldwide heritage conservation, that we owe the first modern foundations of this principle, traces of which have been found for centuries in the world of artistic restoration.[33] In 1939, Brandi founded the Istituto Centrale per il Restauro, the first official school for training in restoration; in 1963 he published

Teoria del restauro (*The Theory of Restoration*). As its title indicates, this work is an attempt at conceptualization that has had very significant repercussions in professional circles and has helped to structure the operational principles that still hold in conservation practices. Discussing both artistic works and monuments, Brandi explicitly opposes scraping and other intensive treatments implemented to combat 'patina'. While he does not argue for the disappearance of all human input, and acknowledges the movement of critical interpretation at work in each conservation initiative, he insists that restoration operations should be exceptional, and that the protection of historic works and monuments should be organized as a mode of conservation conceived as a form of highly controlled intervention. The principle was already mentioned in the Athens Charter and enshrined in the Venice Charter, published in 1964 at the end of the Second International Congress of Architects and Technicians of Historic Monuments. This reaffirms the exceptional nature of restoration (article 9) and declares in its article 4 – the first in the 'conservation' section – that 'it is essential to the conservation of monuments that they be maintained on a permanent basis'.[34]

The expression 'minimal intervention' does not appear in full in the Venice Charter nor in the Burra Charter, which succeeded it in 1979. It was ultimately quite late that it emerged as an important guide in practice, in the form of an *ethical*[35] principle established internationally in the *Management Guidelines for World Cultural Heritage Sites*. The conclusion of chapter 8, a chapter devoted to 'Treatments and authenticity', opens with these words:

> World Heritage is a fragile and non-renewable, irreplaceable resource. The aim of safeguarding World Heritage sites is to maintain their authenticity and the values for which they have been listed. Therefore, any treatment should be based on the strategy of minimum intervention, and incorporate a programme of routine and preventive maintenance.[36]

Minimum intervention, minimal intervention: the formulas are strange and may seem rather comic. But we can also find they have a certain elegance. Almost oxymoronic, they finally underline, without claiming to resolve it, the inevitable dilemma at the very heart of the idea of conservation, which navigates between the concern for physical integrity, the acceptance of the passing of time and the inevitable action of the agent who conserves. Far from appeasing it, the expression acknowledges the anxiety entailed by the activity of care, exacerbated by the constraint of authenticity.

However, its uses have sparked numerous criticisms, which emphasize the fact that the ambiguity of the formula also presents significant risks. These criticisms were voiced during profound upheavals in the field of heritage conservation. Increasing numbers of people started to denounce the Eurocentrism concealed behind the universalist pretensions of the major principles set out in international charters and conveyed by Western restorer-conservators throughout the world. The idea of historical truth, the relationship to the past, the role of humans, the role of nature, the notion of heritage, the form assumed by authenticity: it was impossible to assert that a single path should be imposed on everyone, either in the resolution of these problems, or even in their formulation. It was now vital to open these questions to other actors, to other territories, to other philosophies, some of which, for example, base collective memory on the disappearance of the material traces of things, or do not rely on the need to restrict human intervention – quite the contrary. Let us remember our brief foray into the Shinto temples of Ise, not far from Kyoto, whose conservation involves a complete remaking, repeated over the centuries. It is no coincidence that it was in Japan that this 'culturalist' turn was consolidated in the form of the *Nara Document on Authenticity*, published on the occasion of the conference of the same name organized in 1994 by Unesco, Iccrom and Icomos.[37] Minimal intervention moved centre stage in these debates: it was accused of being one of the Trojan horses of a cultural hegemony that dared not speak its name.

This is not the only reason why the principle has been criticized. It has also been reproached for replacing one positivism with another.[38] Now armed with countless scientific instruments capable of making the smallest bits of matter speak, curators can boast of adjusting their own interventions to the millimetre, so that they almost completely disappear from the landscape. In place of the historical and heroic rationalism of a Viollet-le-Duc, some end up installing a materialist rationalism whose restraint is quite relative. In both cases, it is essentially the myth of maximum objectivity that is being preserved. The interpretive, and always political, gesture of conservation is greatly attenuated, when it is not purely and simply denied. The argument is convincing and encourages us to take the idea of minimal intervention with a grain of salt, at the risk of throwing the baby out with the bathwater and falling into the trap of a questionable humility. As Donna Haraway noted, 'Transparency is a peculiar form of modesty',[39] one that is socially and politically situated, even though it strives to forget this fact. So let's avoid falling into these 'disappearanceist' ways and remain attached to a version of minimal intervention that is not a solution that just needs to be applied to preserve the authenticity of a thing by erasing all traces of human choices and biases but, on the contrary, a problem, a living concern constantly reminding everyone of the ambiguities in the care of things. Seen from this angle, the idea of minimal intervention becomes an operator of material diplomacy that expands the ideas we drew from the work of the Rotor collective in chapter 2.[40] It is no longer simply a question of giving up excessive control over the material and coping with wear and tear, but of taking maintenance activities, from the most commonplace upkeep to heritage conservation, and inventing within them a form of measured action, adjusted on a case-by-case basis, deployed in direct contact with things. This is a matter of finding the most appropriate way of joining in the dance of maintenance, here troubled by the moral burden of authenticity.

Pathways inspired by environmental ethics

The history of debates on heritage conservation, even if far from ending in a universal consensus, shows to what extent maintenance can be a matter of dosage, of a delicate balance to be struck in the very relationship of care. To follow this argument through and grasp the full scope of the problems raised by the subtle adjustment of human action in the care of things, we propose to explore a final avenue, by examining an area where these questions seem to be most purely concentrated: environmental protection. In doing so, we will at first glance go beyond the scope of the maintenance of things. This shift, however, is not a mere detour. Environmental issues and discussions on the relationships between humans and nature, including the questions raised by the simple opposition of these two terms, are essential to questioning maintenance. Mierle Laderman Ukeles weaves close links between domestic, urban and environmental maintenance herself.[41] If the implications of such a rapprochement are far too vast to be tackled in detail here, the question of the place of human action, and more precisely of the forms of intervention carried out in the name of the preservation of the environment, in itself offers an extremely productive entry point, providing a promising first step in this direction and grasping the anthropological dimensions of this aspect of the care of things.

Countless debates have fuelled the environmental question for decades and given rise to a proliferation of theoretical texts, practices and political projects. Without claiming to embrace all the richness of these discussions, it is interesting to note that some of the disputes which have animated the major theoretical models of environmental conservation have evolved in an almost symmetrical way to those experienced in the world of heritage conservation. By emphasizing the linear dynamic of the arguments, we can in fact create a mirror image of the disputes that we have just described regarding historical monuments. Instead of starting from the influence of humans asserting their power over matter to preserve the historical authenticity of things (the model of Viollet-le-Duc), we will here begin with the total absence of humans, claimed as a condition for the preservation of a thing called 'nature'. As the controversies and political and philosophical confrontations evolve, a common territory gradually emerges: that of the measured ways and accommodations through which an act of care is accomplished, one that does not deny its ambivalence, but constantly works to balance it. In this story, the labels are no longer the same. If *conservation* is still very much present, its proponents are no longer opposed to the model of *restoration*, but to that of *preservation*. They no longer call for a withdrawal of humans, but on the contrary demand a form of repopulation.

At the centre of attention we find thus that somewhat bizarre thing, both omnipresent and circumscribed, evident and mysterious: nature. A nature that must be protected. But not just any kind of nature. The discussions that interest us focus on a very particular version of nature, one that emerged in North America in the nineteenth century and is characterized by a certain form of purity, at least concerning the place that humans have in it. Often described as a 'wilderness', it exists without humans. We could even say that it is made up of their absence – an absence that has become the primary condition for its preservation.

As many authors have pointed out, this highly particular nature is closely linked to the very invention of America and its people. With regard to old Europe, it represents a form of 'heritage' for (North) Americans: a world spared from the stains of civilization, the ultimate symbol of freedom, differing in every way from a European history marked by feudal servitudes and ancient slavery.[42] This terrifying nature, which inflicted the greatest trials on the conquerors as they sought to settle ever further west, was at first a source of fear, but in the nineteenth century became the object of a veritable cult in which Henry Thoreau and John Muir were the main devotees. Muir, in particular, a naturalist and philosopher whose accounts of his travels in the north-west of the country have left their mark on entire generations, played a considerable role in designating the truly spiritual virtues of this wild and virginal nature, which he likened to a temple.

Wilderness is therefore at the heart of American identity, even of the American soul.[43] But its celebration developed just as a deep concern about its future was growing. This is the whole paradox of North America: if its evolving population took on political and spiritual substance through contact with the wilderness from which they were able to emerge alive and strengthened, they thereby worked towards its destruction, going so far as to threaten nature with outright disappearance.

Thoreau's writings emphasizing the urgency of preserving nature are well known. But it was undoubtedly Muir, as both theoretician and activist, who had the most important influence in this field by describing the relationship between humans and nature in moral terms. If this gesture seems very exotic from the point of view of the Europe of the Enlightenment, it nevertheless had crucial repercussions in the establishment of the ecological movement and the evolution of the debates which shook it over the decades, first in the United States, then well beyond. Muir also had a great influence on the country's first environmental protection measures, advocating for the creation of Yosemite National Park and founding an environmental association (the Sierra Club) that is still very powerful today. From the first protected national parks to the enactment of the Wilderness Act in 1964, Muir's thinking profoundly influenced US preservationist policy.

However, right from the start, the welcome given to the preservationist model of wilderness was far from unanimous. Gifford Pinchot, a forestry engineer close to John Muir, very early on criticized him for his desire to put nature under a bell jar, preventing any possibility of progress, either technical or social. Pinchot preferred a 'conservationist' model organized around a 'reasonable' management of natural resources that aimed at ensuring that no one, and nothing, would lose out in the process.

It was Aldo Leopold, also a forester, who profoundly transformed the terms of the debate and charted the course for a completely rethought relationship with nature. Fundamentally opposed to the preservationist model, he also distinguished himself from Pinchot's utilitarian project, which, although establishing the principles of balanced management, was essentially dedicated to human development. The radical nature of Leopold's thought lies in the way it generalized the environmentalist position.[44] In his writings, drawing on his own personal experiences, he defends a new ambition, where it is no longer a question of cultivating resources in a reasonable manner, but of maintaining relationships within complex ecosystems rich in their diversity – ecosystems, he makes clear, within which human beings have their place. Along with a few others,[45] John Baird Callicott worked to make Leopold's texts known to as many people as possible and to place them at the centre of contemporary philosophical debates. While extending Leopold's insights, he endeavoured in his own work to continue the exploration of the moral foundations of the relationships between humans and their environment. In an article devoted to the idea of wilderness, Callicott started with a text by Leopold – whose title, 'The Popular Wilderness Fallacy',[46] alone sums up his opposition to the preservationist model – and reassessed the set of problems posed by this very particular definition of nature, a definition that still forms the basis of a large number of ecological programmes.

The confusion generated by the notion of wilderness, explains Callicott, fundamentally lies in the place it leaves for humans. This place is organized around two main ideas. First, it springs from the metaphors of virginity that, in John Muir and Robert Nash, accompany the stories of Christopher Columbus's discovery of a 'New World'. This space is in the image of Paradise: an intact, idyllic site, which has never known the mark of human beings.[47] Nature in its purest state. Callicott insists on reminding us how ethnocentric and racist this idea is. Not only had the lands of what became America been populated for a long time by humans, but the populations who inhabited them were far from being simple hunter-gatherers without any transformative impact on their territory. Today, we know how highly developed were their practices of cultivation and even 'management' of land, forests and rivers.

The second idea that goes with wilderness is a pendant to the first. It is based on the highlighting of the dangers posed by human interventions to nature. The actions of humans are systematically equated with deterioration. From this idea comes a model of preservation that purely and simply erases humans from the picture. If nature must be protected, it is essentially from humans and their main achievement, civilization. The argument comes as a surprise, since it testifies to a complete reversal in the relationship with wild nature, which was understood in the early days of the conquest of the West as fundamentally dangerous and immoral, and whose domestication was meant to ensure a form of redemption for human beings.[48] Still, this vision is very questionable in Callicott's eyes. It condemns in advance any form of civilization, as if it could be nothing other than a vast, uncontrollable operation of mechanization, intrinsically harmful to its environment. Once again, the argument poses considerable problems since it generalizes certain industrial practices to all the features of civilization, ignoring the consequences of such a view for entire parts of the world population.[49]

We can of course recognize in these two ideas, which are two sides of the same coin, the broad features of a 'modernist' ontology of nature whose universal character has been strongly called into question in recent years.[50] Nature is on one side, human beings on the other. Taking care of the first involves removing the second. A considerable number of comments, analyses, definitive opinions and other, more nuanced, views have been published about this separation, and it is important to remember that we are only scratching the surface here. Let us simply note that we can see in this model the result of a sort of alliance between certain religions[51] and the beginnings of modern science.[52] This explains how this ontology of nature is so omnipresent in industrialized countries and why it is still at the heart of many environmental policies. It was this ontology in particular that underpinned the highly influential Wilderness Act passed in the United States in 1964, which 'presupposes, even establishes, the split between humans and nature'.[53]

Several other criticisms have been levelled at preservationist arguments. One aspect is particularly relevant to the question that concerns us, and links up to the debates mentioned previously about certain trends in heritage conservation. Under the seeming obviousness of a maintenance that takes place by withdrawal, in favour of a 'liberated' nature, agency is in fact unevenly distributed. On the one hand, anything that is not human has a very limited capacity for action. The things of which nature is made are as if frozen, demure, reduced to objects of contemplation that humans can enjoy as spectators, from a distance. Callicott repeatedly highlights the glaring limits of these postulates of stability which nourish a vision of nature that can be 'protected' in limited spaces, sheltered from civilization. In the attachment to the wilderness, we can indeed detect elements reminiscent of the quest for permanence described in the previous chapter – natural forms that are intrinsically balanced and which could last for all eternity provided that humans no longer come to disturb them. Such a vision is in complete contradiction with everything that science has taught us, at least since Darwin: 'nature' is constantly transforming, unstable and unpredictable.[54] Conversely, on the other hand, the 'withdrawal' of humans required by the model is quite a relative matter. As Carolyn Merchant shows,[55] there is not a huge difference between the posture of engineers who wish to reproduce a completely controlled nature, and that of environmentalists who, with the help of protected sites, organize the return to a wilderness rid of human parasites. A 'hidden arrogance'[56] lies at work in this form of care where transcendentalism combines with a humanism that sets itself up as the protector of a virginal nature. This false modesty, adds William Cronon, also nourishes a systematic suspicion of all soil workers, in favour of a contemplative relationship with nature which derives from the anti-modern position of the nineteenth-century bourgeoisie, the very people who in reality benefited the most from galloping industrialization. It thus encourages the humans who are most concerned to believe that it is possible to 'escape history' and evade their responsibilities.[57]

Since Leopold, criticisms of the preservationist model have therefore been many and varied. Most point in the same direction: they take into consideration the complex dynamics of relationships at work in ecosystems and hence the impossibility of leaving humans out of the concrete modalities of the care that urgently needs to be given to the Earth. The practical consequences of these principles are essential for understanding the form of maintenance that is being invented here, concerning something as complex as the planet. But they are also essential if we are to begin to understand how each maintenance activity can be evaluated with regard to environmental concerns and fitted into a system of relationships that can no longer be divided into disconnected areas.

By getting rid of nature, the philosophy of Leopold and his successors effectuates a twofold change of scale. Because it highlights the importance of the care given to the 'biotic community', it encourages everyone to raise the environmental question in all generality, while offering suggestions for local activities that are much more nuanced than merely putting nature under a bell jar. This is shown by Leopold in 'The Farmer as a Conservationist',[58] in which he describes the experience of his own family as they set about restoring land that was once cultivated and then abandoned. Beyond its somewhat idyllic aspect, the article has the merit of showing that, if they want to flourish in their environment, the farmer and his family can neither excessively control it nor simply neglect it. Leopold depicts a form of 'relational' care that involves a series of adjusted and informed interventions allowing the environment, including humans, to exist without the whole milieu collapsing for the benefit of one of the entities that inhabit it. We are a long way from the idea of 'sustainable development' as it was later taken up by a whole section of environmental engineering, as a means of creating a consensus between antagonistic logics. This model calls for a much deeper transformation of practices, requiring us to completely abandon the metaphor of nature as a reservoir of natural resources of which we should learn to take advantage to a 'reasonable' degree. Leopold vehemently argues against the industrialization of agriculture and insists on the need to modify agricultural techniques so that the terms of a desirable cohabitation for all can be developed, and human and natural economies can nourish each other.

In recent years, initiatives inspired by this perspective have spread, developing forms of soil maintenance, land cultivation and forest preservation. They all demonstrate in their own way possible forms of 'partnership' between humans, flora and fauna.[59] There are countless projects, and we can mention only a few works that will enable the reader to discover them, from ways of recomposing territories on the scale of 'bioregionalism'[60] to attempts to develop urban agriculture that invent new intricacies between the urban and rural domains,[61] through permaculture techniques[62] or, more modestly, the use of goats and sheep to 'crop' certain green spaces in the city.

In the field of horticulture, the work of Gilles Clément is a fine example of this way of inventing a new, readjusted place for humans in the relationships they cultivate with the living world.[63] The name he gave to his practice, 'gardens in movement', perfectly illustrates the way in which he sees his activity as a landscaper. This is based on the recognition of the dynamic nature of plant life and respect for its unpredictability. Without disappearing completely, the gardener must abandon all pretensions to control and transform themselves into an attentive guide who observes as much as they 'manage' the garden they take care of. It was in Creuse, on unused land, that Clément first experimented with this 'management method aimed at maintaining openness so as to maintain floristic diversity'.[64] Rather than deploying ever more constraints, it is a question of fully recognizing that plants have a right to circulation. By developing this practice, then travelling around the world, Clément not only invented his own gardening methods, but also opened himself up to the diversity of ways that plants inhabit the Earth and move across it. This movement is not 'natural' in the sense of autonomous, but always the result of complex interactions. If the landscape cannot be understood as a fixed and stable form, this is not only because species move, but more generally because it 'has always been subject to modifications, which stem from the climate, from the plants themselves, from humans and animals'.[65] Maintaining a garden in motion therefore does not mean disappearing. It means encouraging the presence of certain plants, removing others, but always allowing a large part of the transformations to take place, taming the fallow land[66] while granting it great freedom. Clément says, for example, that the appearance of a species considered invasive encourages him to transform the path it follows so as to accommodate it while limiting its development. He also explains how the fall of a tree, which prevents mowing, leads to the appearance in the surrounding area of species that have not previously had time to establish themselves. During his intervention at the André Citroën park in Paris, which gave him the opportunity to formulate a manifesto for his model of a garden in motion, his ecological bent became political and showed the main lines of another way of thinking about the presence of plants in the city. He also helped to re-evaluate the profession of municipal gardener.

In forestry, several contemporary examples encourage us to dismantle the clichés that some associate a little too rapidly with the American preservationist movement. In recent years, many ways of taking care of forests have been reinvented that never postulate the need for a total disappearance of humans – quite the contrary. As with Leopold or Clément, what is at stake here is essentially a form of assistance and guidance that does not erase human action but also does not seek to dictate, from start to finish, the ways plants, soils and animals appear, let alone improve them, as if a patch of meadow or a forest were naturally feeble and needed humans to help it be healthy and 'productive'.

The fostering of the pink flamingo population in the Camargue, studied by Raphaël Mathevet and Arnaud Béchet, is a demonstration of the importance of these entanglements.[67] It shows how absurd is the attempt to precisely identify what is at stake – humans or 'nature' – in environmental protection projects. In this case, although there has been talk for years of preserving the pink flamingo in a 'wild' form in the region, this in no way means that the bird is outside the world of human and indeed socio-economic relations. Conversely, the fact that the flamingo has remained in the area, and in particular that couples now breed on certain sites, owes a lot to the interventions of people who have strived to 'manage the territories'[68] for this purpose: a form of action that breaks away from a model of absolute control of the elements without ever completely erasing the presence of humans. The protection evident in the maintenance of certain sites can be thought of in terms of a nature whose multiplicity must be acknowledged and appreciated. These are different forms of naturalness that must be taken into account, within which humans are neither foreigners nor intrinsic disruptors. On the contrary, they can invent the role of 'auxiliaries' for themselves and thereby find 'ways *of belonging to nature* and *engaging* alongside it'.[69]

As we can see from these few examples, it is not a simple technical question that is being discussed and developed here. The issues are systemic. Let's return to environmental ethics. Catherine Larrère shows that it involves an important moral shift,[70] one that enriches our understanding of what the very idea of maintenance, of caring for things, can imply. A first moral consideration of our relationship to nature could in fact consist of establishing all living beings as subjects, which would amount to broadening the scope of morality by adding to human life a long list of 'intrinsic' goods that need to be preserved. This is the perspective of bioethics. But Leopold's philosophy goes much further by opening the way to an 'ecoethics' in which the moral character of each action is judged in terms of its consequences not on this or that being, but on the ecosystem in its entirety. The intrinsic good lies in community and solidarity.[71] Even more than the beings present, it is the relationships that we must learn to make last; these are what should be taken care of.

Ethics and the care of things

We are well aware that it can be mind-boggling to move so quickly between such different worlds. Few things concretely connect the torments of Mustang owners, the doctrine of minimal intervention in heritage conservation, and the philosophical arguments of environmental ethics. By moving from one thing to another, from one type of problem to another, we have somewhat radicalized the principle of peregrination which organizes this book. Some readers will no doubt think that we have gone too far. But remember: our objective is obviously not to come to any conclusions about these different subjects, but rather to show the particular light that is shed on them when we pay attention to the maintenance problems they involve.

We started this chapter by focusing on the question of authenticity. This meant we could extend our questions about the ontological dimensions of maintenance while significantly shifting them. Rather than varying a series of situations in which forms of maintenance are deployed in the service of specific modes of existence, as we did in the previous chapter, we were able to discover variations and gaps within practices that one might initially have imagined were homogeneous. Not everyone agrees on the ways of making something last in the name of its authenticity – far from it. One and the same person can hesitate, adapt, modulate their actions in order to take care of something that they consider authentic. This is because maintenance is a deeply disquieting activity. Understood from the point of view of this disquiet, authenticity appears above all in the form of a worry, a concern – a quality whose contours are never completely predefined, a quality that care practices must continually bring into being. Among the debates that we have mentioned, some quite explicitly flaunt this generative dynamic and assume that maintaining consists of making things exist rather than ensuring their simple perpetuation. However, this active participation in the existence of things can only be carried out within the materials, in contact with the engine parts, the plaster of the cathedral bay, the abundant life of the soil. And because it is always uncertain, this encounter is nourished by doubts and adjustments, improvisations and fluctuations. It is, here too, a matter of dance.

Even though it operates in a regime of authenticity, then, mainte-nance encourages us to cultivate these oscillations, and to embrace the disturbance that they generate. To take care requires *tact*. A material tact which is also an 'ontological tact', following the beautiful expression that Vinciane Despret uses to describe the way in which certain people talk about the relationships they have with the dead who surround them:[72] a tact which does not a priori impose on the things present a dualistic reading immediately placing them on one side or the other of existence. To maintain something tactfully is to compose with the material rather than to try to decide for it, and to cultivate the virtues of hesitation and doubt to this end.

This art of composition requires the steps of the dancers of mainte-nance to be continually adjusted. In many situations, this requires humans to accommodate their interventions. The issue is particularly sensitive in the areas of heritage conservation and environmental preser-vation, where there are intense debates about the place that humans should have in maintenance processes. In both cases a struggle has developed against forms of relational purity: against the idea of things obeying the whim of women and men with all their expertise, on the one hand, and against the principle of the complete withdrawal of humans, on the other. One trend in environmental philosophy opens up some stimulating perspectives by paving the way to a care that is not thought of as necessarily at the service of an isolated 'thing', but is organized around the community and thus around relationships as such. Maintenance, approached from this angle, can be part of *an ethics of coexistence*, which echoes the main lessons of the ethics of care in two ways. First, instead of posing autonomy as an unsurpassable horizon of action, it is entirely oriented towards situations of interdependence, which it takes as its starting point and which it aims to cultivate. And second, faced with the abstract nature of the traditional problems of morality, this form of care favours a material engagement with things, one that is open to the unexpected. This practical maintenance is part of what Maria Puig de la Bellacasa calls a 'situated ethicality' in 'a world constantly done and undone through encounters that accentuate both the attraction of closeness as well as awareness of alterity'.[73]

Finally, we can consider the tact at work in the dance of maintenance as an encouragement to renounce any overly narrow definition of the very idea of action, whether human or not. As Tim Ingold points out, once we take into consideration the variety of elements present, we can hardly limit ourselves to describing a 'dance of agency', in the sense that Andrew Pickering gives to this expression.[74] Taking the example of the kite, Ingold argues that anthropologists should get rid of the overly oriented notion that prevents them from fully understanding the role of air. They must loosen the syntax of their descriptions and become attentive to a 'dance of animacy',[75] in which it is the relationships between numerous partners of different natures which generate the forms and their continuity. If the image is so telling in the case of maintenance, it is because it resonates directly with the concerns of a large number of people involved in the care of things. In a certain way, being tactful and cultivating a situated ethics of care amounts to immersing oneself in a collective dance through which the continuity of things is animated.

That said, Ingold's proposal should not completely distract us from another crucial aspect of the dance of maintenance: not every human participates in it in the same way or with the same intensity. If we must loosen the overly reductive term *making* in the expression 'making things last', we must not forget that it nevertheless describes a task that some people accomplish better than others. Recognizing the importance of situated ethical operations of maintenance requires us to pay attention to those who find themselves in contact with things, directly confronted with the uncertainties of matter – to those who know how to be tactful and who practise the delicate art of hesitation and moderation. This gesture of consideration is all the more important since it is rarely the most visible and the most audible who participate in the maintenance dance. Even if the field of heritage conservation is today saturated with guides, charters and international standards and populated by experts of all kinds who organize maintenance activities, it is still the restoring masons who have the delicate task of *practising* minimal intervention, of finding in a particular situation the right forms of accommodation with the materials of the building, the products available and the instruments they handle.[76] Just as it is up to the operator who 'removes' the graffiti to adjust her or his interventions on site to the reactions of the materials so that at the end of the intervention a facade remains minimally transformed, but cleared as much as possible of the traces of the inscription deemed undesirable.[77]

As regards environmental ethics, the question of knowing who takes care of the Earth is obviously both crucial and baffling. It gave rise to fascinating discussions, particularly around the imperialist implications of the preservationist model associated with the idea of wilderness. Thus, many authors inspired by Leopold's philosophy consider that a form of ecoethics was most likely being carried out on the lands of what would become America well before the arrival of the settlers. We can even legitimately decide that it was thanks to the continued care of indigenous populations that the colonists discovered this rich 'wild' life, which they ended up venerating as a nature without people, systematically denying the contribution of the humans who were fully present in the landscape. However, the stakes of such discussions go far beyond the relationship with America's past, as demonstrated by ecofeminism, which shows on a much larger scale the neglected part of the populations who take care of the Earth. Women and their struggles, particularly in colonized countries, hold a central place in the development of political, artistic and technical actions designed to take into account the relational issues of an ecology of belonging: an ecology that does not aim to separate humans from nature, but to integrate the consequences of human actions on the health and well-being of all within the living environments concerned.[78]

This journey thus leaves us facing a final question, one which we have continued to ponder since our first descriptions of Mierle Laderman Ukeles' performances: *who takes care of things?* The question fuels the power relations which sometimes configure the material diplomacy at work in maintenance. In recent years, conflicts over this issue have multiplied, reminding us that not everyone has access to the negotiating table.

7

Conflicts

At the start of 2020, as the number of patients affected by severe symptoms of Covid-19 increased exponentially and intensive care units came under unprecedented pressure, most hospital systems around the world found themselves caught up in a real storm. Added to the tidal wave of patients was the deep uncertainty regarding the virus, about which little was known at the time, either about the modes of transmission or the potential complications. There was also a shocking shortage of equipment. Gloves, masks, gowns, as well as syringe pumps, beds, oxygen bottles... everything was in short supply, and healthcare workers had to face the worst health crisis in decades by cobbling together makeshift solutions. Several initiatives emerged from civil society to make up for the shortcomings as best as possible. The press reported on sewing collectives working to make fabric masks intended to equip caregivers, and a few groups of 'makers' who organized themselves remotely to design and produce replacement medical equipment as a matter of urgency, wherever possible.

What is less known is that the period was also the scene of a profound crisis of maintenance in the technical departments of hospitals. Against a backdrop of shortages, the available equipment was sorely tested, with certain items operating much more frantically than usual. Malfunctions and breakdowns became ever more common, requiring repeated technical interventions. There is nothing very surprising about that. After all, the machines were being subjected to excessive use, and the technicians who were supposed to take care of them found themselves caught up in the tumultuous activities of the so-called 'essential' occupations, condemned to bear a monumental workload at the height of the crisis. But one thing complicated their task. Some of these machines proved very difficult to repair. This was particularly the case for ventilators, key pieces of the medical system put in place to respond to the pandemic. In Italy, then in the United States and elsewhere in the world, maintainers faced a double constraint. On the technical side, having access neither to repair manuals nor to detailed plans of the devices, they did not have sufficient information to make the settings and adjustments which would have allowed the machines to continue to operate. They were also unable to manufacture spare parts, then in short supply, with the means at hand. These technical constraints were compounded by contractual restrictions. In many cases, the technicians were not accredited by the manufacturers, and the maintenance contracts that linked the latter to the various establishments specified that these local personnel did not have the right to make changes to the ventilators. Each of their operations represented a risk in terms of guarantee and insurance.

The situation quickly attracted the attention of Kyle Wiens, the founder of iFixit, a company specializing in providing documentation and tools dedicated to the repair of objects of all kinds. In March 2020, he set up a project for a participatory database which aimed to make medical equipment repair manuals available to everyone. In May, he published the official announcement of its launch online.[1] The database already contained more than 13,000 equipment manuals from hundreds of different manufacturers, collected with the help of 200 volunteer documentalists. In his announcement, Wiens explained that, while the initiative had initially focused on three central devices supporting Covid-19 patients in hospital (ventilators, anaesthesia systems and instruments for respiratory analysis), it soon widened. iFixit had mobilized many of its own employees to nurture it, and they had contacted more and more doctors, nurses and technicians in the biomedical sector to locate the problems encountered and identify the needs to ensure the maintenance of different machines.

But the initiative was not appreciated by everyone. On 11 June 2020, Kyle Wiens posted on Twitter a letter sent to him by the intellectual property manager of Steris, a medical equipment manufacturer, ordering him to immediately stop publishing his repair manuals or face legal action.[2] Far from being intimidated, and already accustomed to this type of threat, Wiens refused to comply and sent a letter co-signed by the Electronic Frontier Foundation, in which it was noted that as a host site, iFixit was protected by the law of its country (in this case, section 512 of the Digital Millennium Copyright Act) and could not be held responsible for content provided by its users.

In parallel with the launch of the iFixit database, the US Public Interest Research Group published a petition addressed to ventilator manufacturers and signed by more than 43,000 people. This requested that all restrictions on maintenance interventions be lifted and that information useful for on-site repairs be made available to all, in order to help healthcare establishments care for their patients during the crisis.[3] The Californian branch of the Group had sent an official letter to legislators in Congress, signed by 326 expert hospital technicians, demanding access to all troubleshooting materials (all information, all software, all spare parts and tools necessary to carry out corrective and preventive maintenance actions in accordance with the recommendations of manufacturers, such as repair documentation, diagrams and diagnostics).[4] After all these demands and large-scale participatory initiatives, many manufacturers (but not all) agreed to lift the restrictions for a time and not to sue people who shared the information necessary to repair their machines or those who carried out their maintenance outside the conditions provided for by commercial contracts.

This maintenance crisis in times of pandemic is very revelatory. It bears witness to an unprecedented time during which maintenance and attempts to hamper it came to the fore, while prolonging the operation of machines became a matter of life and death. In a hospital, the line between caring for things and caring for people simply does not hold. Both are closely connected. By accentuating forms of wear and increasing the number of breakdowns, the compressed time of the pandemic led to a pile-up of maintenance activities and considerably accelerated their pace. But while it revealed the crucial role of the maintenance of hospital equipment, the health crisis also brought to light the restrictions to which maintenance activities may be subject, and the tensions they arouse. The constraints faced by hospital technical staff, the various initiatives to distribute manuals, and the initial reactions of the manufacturers revealed a struggle that focused on a twofold question that is much less trivial than it seems: who must, and who is able to, maintain the relevant machines?

In the specific case of medical equipment, two main positions clash: that of the large industrial groups which impose the technical and legal conditions for the maintenance of the objects they manufacture and sell; and that of 'local' technicians, in direct contact with the users of these objects, who resist these conditions and invent numerous ways to circumvent the restrictions. Looking at the terms of the debate, we can see that in fact, while the pandemic accentuated tensions and helped to make the struggle public, the situation was not new. This is partly why the iFixit operation was so successful. For several years, collections of manuals, scanned or retrieved online, had been circulating within the community of medical technicians, where they were surreptitiously exchanged. Kyle Wiens' initiative was itself inspired by a database hosted in Tanzania, Frank's Hospital Workshop,[5] assembled and updated by Frank Weithoner, a biomedical engineer who had set himself the goal of facilitating the work of his colleagues in the Global South, where machines purchased second-hand are numerous and the maintenance contracts offered by manufacturers are often untenable. Conversely, in the United States, manufacturers in the field had long been aggressively lobbying in favour of strengthening restrictions. The Advanced Medical Technology Association, among others, with more than 400 medical equipment manufacturers (including major bodies such as Siemens, GE Healthcare and Philips), regularly pleads with legislators to reject any relaxation of the conditions of maintenance of the machines they sell, and for very close control of on-site interventions. Likewise, the Medical Imaging and Technology Alliance, which represents around fifty companies in the field, constantly lobbies the Food and Drug Administration (FDA) in order to further reduce maintenance technicians' room for manoeuvre. It points in particular to the risks that repairs represent, and strives to have them viewed as 'remanufacturing' so as to widen the scope of its arguments. The issue at stake is an important one: all these manufacturers depend on the FDA, and if their machines were to cause the death of a patient, they would be legally held responsible. On this question, the FDA sought to explicitly reassure them by publishing a report in 2018 in which it questioned their pressing requests for an additional strengthening of restrictions. It noted that the maintenance operations carried out by hospital technicians do not present any danger and are, on the contrary, essential to the proper functioning of establishments and the safety of patients.[6] Aside from questions of safety and liability, as documents published by the Medical Imaging and Technology Alliance clearly explain, manufacturers also seriously fear for their brand image. They refuse to allow their machines to be subject to tinkering and ad hoc transformations that they cannot control. Despite the FDA report, the fight has not stopped and lobbying actions continue – especially since, quite apart from the pandemic and the case of medical equipment, the question of *who* has the right to carry out maintenance operations on *what* has become increasingly sensitive in many sectors.

The conflicts around the maintenance of ventilators in times of pandemic mark an additional stage in the already rich history of a struggle conducted by activists in North America and Europe to defend what has gradually stabilized around the formula the 'right to repair'. The battle dates back to the beginning of the twentieth century and has moved from object to object. Among these, cars played an important role as early as the 1910s, when Ford tried to restrict the repair market to ensure the longevity of his Model T by creating a network of authorized garages and putting tools and 'proprietary' parts into circulation. The initiative did not take off, and the standardization of the maintenance of Ford cars ended up being abandoned.[7] But the episode marks the beginning of a long history of contentious relations between automakers and the independent repair community. It was when these relations became tense again, as electronic maintenance instruments appeared at the end of the twentieth century, that the idea of a 'right to repair' was explicitly formulated by mechanics' associations. They fought to obtain the possibility of using this new equipment against the wishes of the manufacturers.[8]

Meanwhile, the IT sector has also been the scene of tensions linked to attempts to control repair activities. In 1956, IBM, then largely dominant in the market for the first big computer mainframes, was sued under antitrust law for attempting to limit repairs by people outside the corporation.[9] In order to avoid prosecution, its leaders abandoned the monopoly position, agreeing to train technicians from elsewhere, making spare parts available and disseminating the information necessary for the maintenance of its machines. In just a few years, the sector started to flourish, and the proliferation of repair professionals encouraged the emergence of a significant second-hand market.

Later, it was personal computers and smartphones that found themselves at the centre of heated debates, particularly around the first publications of the iFixit site, created after Kyle Wiens' restrained attempts to put repair manuals for Apple products online. In a few years, sometimes by mechanically preventing its machines from being repaired outside the circuits it controls, the company became one of the most emblematic enemies of a resistance movement which gained media visibility from the 2000s onwards. This era saw the creation in North America, then in Europe, of different collectives representing consumers and independent repairers (such as the Repair Association), and public interest groups (the 'Right to Repair' section of the US Public Interest Research Group).

In recent years, the agricultural world has found itself in the spotlight following several articles and reports, initiated by the online media platform *Vice*,[10] which introduced a wide audience to the difficulties encountered by owners of John Deere tractors. Their machines, increasingly sophisticated, and driven by software that 'protects' them against any intervention carried out outside the network of company technicians, can no longer be repaired as they had been previously, that is to say on-site, by farm personnel. Subjecting farmers to unsustainable prices and deadlines, the situation gave rise to legal action, but also to circumvention practices which led some farmers to become 'hackers' of their own tractors.

Over the years, an international movement has formed to defend a right to repair whose scope has gradually expanded, while the objects affected by the restrictions have multiplied. The action of these activists is now beginning to bear fruit and the balance is shifting, in particular because some of them have succeeded in having their demands transformed into legal concerns. In Europe, things have recently accelerated, and have even led to the inclusion, in the new action plan for the circular economy of the European Union, of a prolongation of the lifespan of consumer goods,[11] in addition to traditional actions in the field (such as reducing energy consumption and improving recycling possibilities). This new objective is explicitly associated with an obligation on the part of manufacturers to promote the 'reparability' of their products. In the United States, although the movement is older, it was slow to consolidate. In 2019, around twenty states discussed bills establishing a right to repair for consumers and third-party services, without success. As initiatives multiplied and more and more public figures called for these laws to become general, lobbying from large groups of manufacturers in many sectors strengthened. Since then, apparently important steps have been taken. In July 2021, Joe Biden signed an Executive Order calling on the Federal Trade Commission to outline regulations preventing manufacturers from restricting maintenance and repair practices performed by consumers or independent technicians.[12] In November 2021, Apple representatives announced, to everyone's surprise, the launch of a 'self-service'[13] repair programme which, from 2022, would allow anyone to purchase replacement parts and consult repair manuals for the brand's recent products.

What exactly is at stake in these battles to determine who is able to maintain this or that object? What can we learn from the resistance movement which advocates a right to repair and, more generally, from the conflicts which run through this side of the care of things? If it seems important to us to end our discussion by looking at these struggles, it is because they extend, and sometimes shift, many of the questions we have addressed so far. Calls to establish a right to repair, in particular, demonstrate a strong desire to promote maintenance activities. The explicit demand to practise these activities contradicts the idea that these practices are doomed to be poorly considered, inevitably relegated to the status of 'dirty work' and mostly rendered invisible. Here, the possibility of making things last is a matter of a redistribution of powers, something that makes it possible to establish a new balance of forces in the organization of the consumer market. This balance of power can of course be reduced to the collectives that it involves – industrialists, independent professionals, consumer associations, activists – but this would amount to erasing from the debates much of what makes them so rich. It is not just a matter of struggles between actors with pre-established representations and interests, but also of forms of the relationship to things that are set against one another, developing in parallel and being redefined through their confrontations.

To follow this last line of inquiry, we must return to the question of temporality. In chapter 5, when we sought to distinguish ways of making things last by showing that time could be problematized in different ways, we started with the idea of *prolongation*. We presented it as the most banal and least spectacular form, since it makes time work only in the light of an imprecise, almost daily projection, which appears at first glance to be quite unambitious. Moreover, in most cases, it is very mundane objects that are affected by this form of maintenance, familiar

THE CARE OF THINGS

to everybody. This maintenance is simply meant to ensure that they last 'a little longer'. However, it is precisely within this framework that the struggles for the right to repair take place. They concern manufactured goods and make relatively modest demands in terms of temporality. There is no question here of developing devices that ensure controlled lifespans or optimization procedures that describe in detail the principles of preservation and precise methods of intervention. All that is being considered, and gradually implemented in law, is the *possibility* of making objects last a little longer than they last today. This plea for prolongation reminds us that, far from being a stable and indisputable fact, the lifespan of objects, however banal they may be, is also a political matter.

Shortening the life of goods

The expression that best crystallizes the political history of the duration of consumer objects is undoubtedly that of 'planned obsolescence'. In recent years, it has also enjoyed much greater success in the French media and political space than the notion of the right to repair. Like the latter, it has accompanied conflicts pitting part of the population against large industrial groups. In France, it has lain at the heart of very specific controversies, which have long limited its political impact, as Jeanne Guien shows.[14] Schematically, the idea of planned obsolescence is used to describe design practices that aim to reduce the lifespan of certain objects. It is criticized as a form of artificial encouragement to consume more, one whose environmental impact is extremely negative. However, for years, French debates have been reduced to a question that seems a little absurd, on the face of it: do these practices really exist? Articles, editorials, documentaries and entire books have been devoted to this somewhat simplistic question. This questioning, in the form of a doubt, constituted the main, if not the only, argument deployed by those who sought to minimize the alarm expressed by critics of planned obsolescence. We could summarize their point of view this way: admittedly, reducing the lifespan of objects would be an economic, ecological, even moral scandal, but the concerns are unfounded since such practices do not exist. Planned obsolescence is a myth. It would be difficult to pose a debate in less constructive terms.

A simple look at the literature in economics and marketing produced for several decades shows how incongruous this French controversy is. The position of the critics who present planned obsolescence as a fable is quite untenable. Not only has the possibility, even the necessity, of reducing the lifespan of consumer goods been discussed at length among market professionals, but it has also been the subject of explicit and recognized implementation in several domains. It also found itself, very early on, at the centre of intense debates and arguments. Retracing this history in broad strokes will help us to better understand what has been happening around the control of the lifespan of things for more than a century, in the context of mass consumption.[15] We will discover that this question has gradually been reconfigured. To put it simply, while at the beginning of the twentieth century the ability of things to last became a problem for some people who saw it as a brake on economic growth, today this durability is seen as one of the ways of reducing the environmental footprint of a human life that has become sick from overconsumption.

Traces of a strategy aimed at limiting the lifespan of manufactured objects can be found in many areas. In 2014, Markus Krajewski described in detail an episode in industrial history that constitutes a canonical example: efforts to reduce the average operating time of electric light bulbs.[16] These efforts were organized within the Phoebus cartel, an oligopoly bringing together hundreds of companies and including in its ranks the giants Philips, Osram and General Electric. Founded in 1924, the organization aimed to control the manufacture and sale of incandescent light bulbs and to protect its members from competing technologies entering the market. Although it failed in this last task, the cartel nevertheless succeeded in structuring, for a time, a market whose prices remained high while production costs continued to fall. Above all, it initiated major engineering work to reduce the lifespan of bulbs. It then required each of the companies that joined it to adopt these standards and not to market any bulb capable of lasting longer than the average time established by the organization. Accompanied by centralized monthly tests, the strategy bore fruit. In fifteen years, the average operating time of bulbs produced by Phoebus members' factories was reduced by a third, from 1,800 to 1,205 hours.

This case alone offers a perfect illustration of the way in which certain actors have used the lifespan of consumer products as an adjustable parameter in market mechanisms.[17] It helps us understand that the idea

THE CARE OF THINGS

of prolongation and, more generally, the principle of wanting to make things last is not based on universal values with incontestable virtues. The words of Anton Philips, then head of the company of the same name, are particularly enlightening from this point of view. Philips complained to one of the directors of General Electric that some of the models they manufactured offered a lifespan much longer than that required by the cartel's conventions. He explains:

> This, you will agree with me, is a very dangerous practice and is having a most detrimental influence on the total turnover of the Phoebus Parties. [...] After the very strenuous efforts we made to emerge from a period of long life lamps, it is of the greatest importance that we do not sink back into the same mire by [...] supplying lamps that will have a very prolonged life.[18]

The terms of the problem are clear: it is not so much a question of reducing the lifespan of the bulbs available on the market as of not extending it. By pushing the rhetoric to the limit, and forgetting the efforts put in place to limit the operating time current before the intervention of the cartel, we could almost compare it to the concerns that we mentioned in the previous chapter about certain preservation practices that have been considered too interventionist. Let's not overdo it, Philips seems to be asking: he sees a 'long' lifespan as an artifice, even a defect, rather than one of the distinctive qualities of the products that he and his colleagues are offering for sale.

If it is important to identify the logic at work in Phoebus's initiative, it is because it is far from isolated. It is no exaggeration to say that the beginning of the twentieth century saw the rise of an army of professionals of all kinds determined to wage a war against the longevity of consumer products, or more precisely to pave the way of 'mass consumption' by fighting against the longevity of products.[19] Given the typical concerns of our age, it is difficult to assess the extent of this war. However, it was a spectacular affair. It started indirectly with the invention of disposable products, particularly in the field of male hygiene (with Gillette razors) and female hygiene (with Kotex sanitary towels), before expanding considerably in many sectors of consumption. The marketing of these objects which had no vocation to last was completely unprecedented in such huge proportions, and explicitly based on a denigration of the ordinary practices of maintenance, presented as a burden from which consumers could free themselves. 'No stropping. No honing', boasted Gillette's advertising for its 'safety razor': the blade no longer needed to be sharpened, just replaced. This masterstroke was made possible by a succession of innovations, both in the metal industry and in the cotton and paper industries, which led to the production of objects that demonstrably cost less to replace than to maintain.

It was a spectacular revolution in consumption practices that emerged during this period, first in the United States, then in Europe. In a few years, arts of making that had been rooted in longstanding peasant and household traditions were relegated to the margins and explicitly devalued, when not frankly stigmatized. Repairing, mending, patching, reusing: the era swept away what Susan Strasser calls a 'stewardship culture' of objects in favour of a 'throwaway culture' which quickly established itself and continues widely today.[20] Taking care of things had become superfluous and, what is worse, suspicious. Cities and newspapers were invaded by messages extolling the moral and civic virtues of consumption and pointing an accusing finger at anything that could be related to thrift. The hour of abundance had come. Men, and even more so women, from towns and the countryside had to play their part in the same way as engineers and their ever more promising innovations. People had to abandon objects that were too old, stop cultivating the talents of prolongation, and throw themselves headlong into the joys of consumption. 'Beware of Thrift and Unwise Economy', warned posters put up in 1917 in thousands of shops across the country.[21] In 1921, the campaign reached its peak, as New York retail businesses founded a National Prosperity Committee to combat all forms of frugality. The injunctions were unequivocal: buy, buy now! The economic health of the country is at stake. Your very morality is at stake. Every time you make something last rather than buying a new one, you're being unpatriotic… These injunctions promoted a certain style of consumption, able to free itself from the constraints of maintenance and reuse. The promise was that of a carefree attitude: it was now possible to buy things, use them and throw them away, without worrying about objects beyond the now very narrow perimeter of consumption. In this vast movement, which saw the birth of what has come to be known as 'consumerism', it was affirmed that it was no longer necessary either to 'care about', or to 'take care of', a considerable number of things.[22] And both these carefree attitudes were presented as a liberation.

With the invention of disposable products and the systematic denigration of domestic practices of upkeep and maintenance, the question of the lifespan of objects came very explicitly to occupy the heart of both the technological and commercial innovations which characterized the advent of the 'consumer society'. It was even one of the main drivers in the process, a pivotal element in the simultaneous configuration of supply and demand by emerging professions specializing in shaping markets: advertisers and experts in what was soon to become the aptly named business of 'marketing'.[23]

The trend quickly grew, until it crystallized in a sector that seemed the polar opposite of inexpensive disposable products and the culture of systematic replacement: the automobile. The episode is well known to business school students and is one of the most famous myths in the history of management and market innovations. It took place throughout the 1920s and 1930s. In the narrative canons of the genre, it is presented as setting up two heroic figures against one another: Henry Ford, of the eponymous company, and Alfred Sloan, then at General Motors. It is also customary to see it as a real turning point in the automobile industry, even in the history of consumer markets. It all started at Ford with the Model T, renowned for changing the scale of the automobile market in the United States, making the car a mass consumption product. Relying on new production principles and instruments, standardized to the extreme, Ford marketed a car whose price was much lower than those of its competitors. In doing so, the company won over a considerable number of customers who previously could not afford a personal vehicle. To win this gigantic market, the Model T also had to present a crucial quality in Ford's eyes: it had to be robust. If emerging middle-class households could be tempted to buy this car, it was not only because it was inexpensive, but also because its longevity was foolproof. This first purchase also promised to be the last.

In his autobiography, Ford expresses this strategy very clearly:

> It is extraordinary how firmly rooted is the notion that business – continuous selling – depends not on satisfying the customer once and for all, but on first getting his money for one article and then persuading him he ought to buy a new and different one. The plan which I then had in the back of my head

[…] was that, when a model was settled upon then every improvement on that model should be interchangeable with the old model, so that a car should never get out of date. […] We want the man who buys one of our products never to have to buy another.[24]

It was precisely this commercial principle that Sloan shattered as he rose to the highest positions at General Motors. Steeped in a techno-evolutionist ideology, Sloan was not satisfied with the image of an automobile market that relied on technological stability instead of embracing the promise of progress. For this was indeed the consequence of Ford's strategy: the quotation above was followed by a sentence which claimed the need to freeze the technical characteristics of its models: 'We never make an improvement that renders any previous model obsolete.'[25] For Sloan, on the other hand, any idea of the status quo was nonsense, for two reasons. First, on a technical level, it was impossible to imagine that with the frenetic pace of innovations driving the start of the century, the automobile industry would fail to improve its products. Second, it was commercial nonsense, which threatened to lead the entire sector into a dead end: once everyone had bought a car for their entire life, how could manufacturers survive? So that the market would not soon wither away, Sloan concluded that it was necessary to fully embrace the virtues of technological progress and rely, on the contrary, on the possibility of making each model progressively obsolete in order to promote a logic of replacement.

As is well known, Sloan – investing first, as a good engineer should, in functional improvements alone – very quickly generalized this logic by decoupling it from technological aspects. It was by focusing on the visual characteristics of automobiles that he was able to fully configure a market in which a changing supply regularly encounters a demand eager for novelty. The challenge was first to win back customers from Ford, something he was able to do with the 1923 Chevrolet, with its completely new outline, imitating that of luxury cars. But it was above all by striving to make models within his own brand obsolete that Sloan tested his strategy to the full, going so far as to take inspiration from the world of fashion and releasing a new 'collection' each year. The models were given the year of their release, and the features of each vintage also signalled a 'move upmarket' supposed to accompany the salary increases of their owners. In contrast to Ford's initial principles, and in line with the transformations that accompanied the development of disposable products, novelty and change had become essential mechanisms of the automobile market. Here as elsewhere, advertising played a fundamental role, becoming an important expenditure item, just like the equally strategic area of design.

Duration as a problem, obsolescence as a solution: this is how we could, without too much exaggeration, summarize the equation which accompanied the structuring of a mass consumption market, freed from the risks of saturation. Over the years, the idea that obsolescence is not a 'natural' material property, but a quality that can be adjusted on the supply side itself, has become established almost everywhere, praised by several economists and intellectuals as the key to prosperity. As Paul M. Mazur put it in 1928, 'wear and tear alone made replacement too slow for the needs of American industry. Business gurus then elected a new goddess to take her place [...]. Obsolescence has become sovereign.'[26] The crisis of 1929 did not slow down this trend, quite the contrary. Some saw controlling the lifespan of objects as the key to relaunching the economy. The idea is notably defended in *Ending the Depression through Planned Obsolescence*, a pamphlet which achieved great success: its author, Bernard London, calls for obsolescence to be organized at state level. The latter would be entrusted with the possibility of decreeing the death of certain products (which would then be purely and simply destroyed) in order to ensure that the machine of consumption resumed unstoppably, with a return to full employment. Without going that far (although some standardization efforts in various fields may be similar to this logic), investment in numerous forms of control (technical, cosmetic, psychological) of the lifespan of consumer objects became commonplace. From the 1970s onwards, several economists even strengthened the theoretical description of the model based on the Coase theorem, which specifies the mechanisms of monopolies based on the sale of highly durable goods.[27]

The values of duration

Mass consumption and the control of obsolescence have therefore become consubstantial in the minds of many entrepreneurs and in the very arrangement of a form of market with hitherto unprecedented growth. Over the course of this industrial and commercial history, however, this apparent unanimity has broken down. As early as the 1950s, the 'manipulation' of the lifespan of objects found itself at the centre of vehement critiques, expressing despair that limiting the durability of products was considered an essential condition for prosperity. One of the central figures of this sceptical movement was Vance Packard, who published two successful books whose titles sum up his position well: *The Hidden Persuaders* and *The Waste Makers*.[28] In them, he develops a detailed analysis of the logics of obsolescence implemented by manufacturers and their advertising accomplices, which he considers to be at the heart of a profound change in values in American society. It is the model of overconsumption as a whole that must be called into question, asserts Packard, the 'throwaway culture' on which it is based and, more generally, the new propaganda of advertising and marketing professionals. Packard compares the situation to the fictions of Aldous Huxley and sees in the artificial production of obsolescence a symptom of civilizational decadence. His books gave rise to violent reactions from theorists who insisted on the economic efficiency of planned obsolescence. But they simultaneously opened the way to a more general critique of advertising tyranny and the manipulation of desires, which can be found in the work of Marshall McLuhan and Jean Baudrillard's earliest texts.[29] Packard's arguments also outline the beginnings of a movement which grew in the 1960s, to the point of organizing itself into a true counterculture. Within this movement, resistance to overconsumption, the refusal to buy new products, and recycling practices inspired by peasant traditions have played an essential role.

In the 1970s, this critique joined forces with emerging ecological concerns. Gradually, Packard's 'waste' found itself problematized as rubbish, residues, and therefore as pollution. When understood in terms of the artefacts and abandoned materials it generated, the culture of obsolescence appeared in a new light, while the cycle linking industrial production and mass consumption was reassessed with regard to its environmental footprint. The work of Barry Commoner played an important role in this new problematization. He advocated questioning technical progress with regard to the pollution it generated and systematically measuring the ecological impact of innovations.[30] By redefining the problem of obsolescence, Commoner introduced 'the awareness of a "biochemical" problem into the heart of what was primarily an economic, technical or social problem'.[31]

With this ecological critique, of which we also find traces in the famous report of the Club of Rome,[32] the obsolescence of consumer goods no longer appears as a solution to the problem of how to boost consumer markets, nor even as an artificial commercial technique with dubious moral consequences. It becomes a problem of sustainability. And if the point is still to act on the lifespan of products, this is no longer to reduce this lifespan, but on the contrary to try to extend it. The register of prolongation, discredited at the beginning of the twentieth century, is once again valued positively. In this shift, making this last is no longer a matter of practical common sense confined to the ordinary household economy, but part of a global strategy that plays a part in preserving the environment on a planetary scale, by mechanically limiting the volume of waste while slowing down overproduction and reducing pressure on resources.

However, it was many years before this question of the lifespan of consumer goods became a salient problem in articulating the economy and ecology. The first emergent legal and normative constraints which were gradually organized on the 'circular economy' model were concentrated in three directions: the reduction of energy consumption, the improvement of waste management, and the development of solutions of recycling. These three areas, all of them pillars of the ambiguous notion of 'sustainable development', result from a delicate consensus which many people consider to have only minimally engaged industrialists, in particular because they have not disturbed the economic models which are still largely based on a logic of systematic replacement.[33]

In recent years, concerns and debates have moved on. In the United States, while the first demands of right-to-repair activists emphasized the question of consumer freedom, they gradually became linked to environmental issues. At the second conference of the Maintainers community, then hosted by Andrew Russell and Lee Vinsel,[34] Kyle Wiens, the founder of iFixit, told us that it had taken him a little time to realize to what extent individuals' practices of maintenance could impact on the environmental footprint of consumption in a much more significant way than recycling, which he now judged to be an ambiguous activity.

In France, the movement was first consolidated around the issue of planned obsolescence. In 2015, the practice became illegal under the energy transition law, for the first time anywhere in the world. In Europe, the contours of the circular economy were extended to the question of the lifespan of products, particularly in line with the work of the French Environment and Energy Management Agency, which in 2014 made 'extending the duration of use' one of the principles of the field, and explicitly mentioned repair as one of its practical concerns.[35] In 2017, the European Parliament published a report which resulted in a proposed resolution.[36] Since 2018, several European Union resolutions have mentioned the lifespan of objects and referred to the repairability of consumer goods. In March 2020, the term was officially integrated into the definition of the circular economy adopted by the European Commission on the occasion of the publication of its new plan for the circular economy. Finally, in April 2024, the European Parliament adopted a directive on the 'right to repair' for consumers, which has been celebrated by Right to Repair Europe, a coalition representing 130 organizations from more than 23 European countries, whose campaign was pivotal in bringing this law to fruition.[37]

These developments mark a double shift. First, they confirm that the problematization of the lifespan of consumer goods has been inverted. Public institutions are now investing in systems that foster initiatives for prolongation that have long been discouraged or even prevented. And large companies, such as Apple, Microsoft and Seb, seem to be following suit.[38] Second, this comes with a significant broadening of the range of people who can once again interfere in the life of things, at the forefront of which are consumers, who can escape the position of passivity to which many industrialists have sought to confine them for years, and can now cultivate new relationships with the goods they acquire. This is an essential argument, one taken up by many activists since the movement consolidated at the end of the 1990s: the right to repair is not only a redefinition of the material properties of objects circulating on the markets, it also concerns – first and foremost, in fact – those who buy them. It is above all a matter of emancipation.

The emancipation of use

From the perspective of the person buying something (a smartphone, a toaster, a tractor), the question of whether or not they have the right to repair the item concerned has some pretty significant ramifications. It has, in particular, repercussions on one's own status as a user, since the right to repair poses the generic problem of the scope of possible actions that a person is able to carry out on property that belongs to them. Discussing the limits of this right amounts, in this perspective, to putting into debate certain aspects of an endlessly tantalizing question: 'What is use?'

The answer is far from unanimous, even in the academic world. It would be pointless to delve into the multitude of studies undertaken on this subject, but we can note that the place of practices of maintenance in the sociology of use, as in the anthropology of consumption, is not self-evident. As elsewhere, these practices generally go unnoticed and are not understood as playing any part in the cultural dynamics that most studies seek to highlight, including when they question the material aspect of these dynamics, apart from their symbolic dimensions alone. Mending, repairing, maintaining, are not really ways of using things, much less of 'consuming' them. However, once we observe consumer objects in situ, that is to say in the daily life of homes, it becomes impossible to clearly separate what is use from what is maintenance. Nicky Gregson and her colleagues bring this out by inviting us to follow the life of a dining room table, carefully restored by the father of the family; of a TV cabinet which, after an unfortunate accident, is tinkered with by the man of the house before finally being replaced by his partner (she does not deem the adjustments made to be worthy of their household); and even a sofa that is really difficult to maintain with a baby around.[39] Consumption does not stop at the act of purchasing, and it is a shame to exclude from the analysis of 'uses' the countless more or less daring initiatives that play a part in the fate of objects well beyond market exchange.

Right-to-repair activists explicitly raise the question of the place of maintenance in consumption themselves. From the beginning of the 2000s, their first media positions focused on the problem of the room left to the user once the purchase was concluded. Very soon, they took their ideas a step further by stating that the struggles waged against the companies which design, produce and sell the goods concerned should lead to a reconsideration of the very idea of ownership. The latter, they assert, has been progressively and insidiously undermined by the various restrictions put in place by big commercial groups. And, as the opening lines of the iFixit Repair Manifesto state in large letters: 'If you can't fix it, you don't own it.'[40]

Thus, the arguments of right-to-repair activists are valid as much for their practical demands as for the more general view they take of the evolution of the 'consumer society'. The difficulties that some consumers encounter in extending the lifespan of objects must, in their eyes, be placed within a broader movement. A movement that has resulted in the decline of consumers' property rights, while the right of manufacturers has stretched beyond the quite unequivocal boundary represented by placing something on the market and then selling it. It is not a simple 'impediment' to use that is denounced here, but a much deeper tendency of many manufacturers to seek to keep control over the destiny of objects after the moment of their acquisition.

These attempts are not entirely new, as we have seen with Ford or IBM, which for a time aspired to maintain control over the 'after-sales' of their products by verticalizing repair interventions. But these initiatives, which ended up collapsing, are incommensurate with the radical trans-formations which started to be felt in the world of mass consumption with the gradual electronization of objects, then their 'digitalization'. This consisted of equipping numerous everyday consumer products with chips and other digital components that play a more or less essential role. Much has been said about this trend, which has complicated many objects whose technical functioning has become difficult to understand at first glance. But conflicts over the right to repair have helped to bring to light a previously unnoticed aspect of the massive digitalization of consumer goods: the significant reconfiguration of the scope of their ownership. Each component using software code operates, in effect, like a Trojan horse which by its simple presence subjects the products thus 'occupied' to the rules of intellectual property.[41] This is how many activities that were once part of normal, private use now come under laws that undermine their legitimacy.

Intellectual property is a nodal point in the struggles between consumers, independent repairers and manufacturers. Even before the wave of digitalization, it had been used to hinder Kyle Wiens' first initia-tives aimed at putting Apple's repair manuals online. Instructions for use, technical plans, repair manuals: all these documents, designed to facilitate maintenance operations, constitute texts which can be protected simply

THE CARE OF THINGS

by intellectual property law. They are still protected and, in certain areas, can be obtained only on a black market where they circulate at high prices. But the scope of intellectual property in the field of consumption is much greater today, since it has been able, through software code, to infiltrate even the material interstices of technical objects.

The roots of these radical transformations go back to 1998, when the Digital Millennium Copyright Act was passed in the United States. This well-known law was initially intended to combat digital piracy. Because it views any attempt to circumvent a protected operating system (for example by a password) as amounting to violating copyright, the DMCA, initially reserved for a limited domain, has led to a tremendous explosion in copyright. This latter was able to extend its action well beyond the world of software strictly speaking, as fragments of code were integrated into the most ordinary objects and established as legal locks intended to 'protect' consumer goods from their own users.

This is exactly how farmers found themselves prevented from repairing their shiny, overpriced John Deere tractors. In the mid-2000s, many owners found it technically impossible to repair their machines. Protection systems had been implemented within them in order purely and simply to block the tractor from starting if an 'external' intervention was carried out.[42] Only one solution, therefore, remained to deal with a breakdown or make an adjustment: to go through the official circuit controlled by the manufacturer. The dream of verticalization of IBM and Ford was finally realized, linked to the mechanics of the thing itself and covered by intellectual property rights. But, in addition to the considerable cost of the operations, the farmers found themselves at the mercy of the rhythm of an official repair circuit that was sometimes very far from their workplace, and soon involved waiting times incompatible with the smooth running of agricultural work. A breakdown, however innocuous, could turn into a disaster by immobilizing the machines for the entire, non-extendable time of a harvest. Suddenly, farmers found themselves dispossessed of their ability to act, unable to do what they and their families had always done: take care of their machines.

Even more than the dramatic situation in which they found themselves, it was the solutions that the farmers found to circumvent the constraints which promoted this issue as a prominent high-profile media case, helping to raise awareness of the struggles for the right to repair. In order to still be able to carry out the interventions themselves, the farmers had to hack their own tractors, in the literal sense of the term, by finding 'cracked' versions of the control software via clandestine networks (notably Polish and Ukrainian). In doing so, they became criminals under the law, contravening the basic rules of intellectual property rights.

The matter proved far from merely a local issue. In parallel with the development of farmers' collectives, organized on the open-source model,[43] it gave rise to a lobbying campaign by Wiens and his comrades which ended up bearing fruit. In 2015, the Library of Congress made an important gesture towards the right to repair by agreeing to introduce an exception to the Digital Millennium Copyright Act for land vehicles, including tractors, which allows modifications to their control software where 'the circumvention is a necessary step taken by the authorized owner of the vehicle to enable the diagnosis, legitimate repair or modification of a function of the vehicle'. However, this was a very limited victory. The concrete conditions of the exception are reduced to a very small number of situations. Above all, John Deere, the tractor brand that caused the scandal, did not comply. Since 2017, it has made buyers sign a clause which again prohibits them from modifying the software and making modification to their tractors.

This case is a perfect illustration of what software does to the ownership of consumer goods and what the question of the right to repair makes evident in its transformation. Through computerization, the status of the object has been profoundly modified while its ownership has been dislocated and redistributed:

> Dave paid for the tractor; he owns what's tangible: the wheels, the metal chassis, the gears and pistons in the engine. But John Deere owns everything else: the programming that propels the tractor, the software that calibrates the engine, the information necessary to fix it. So, who really owns that tractor?[44]

We can see that the struggles waged by certain actors so that people 'themselves' can play a part in making things last go beyond the apparent banality of the material operations to which they are nevertheless firmly attached. They give a completely different twist to the innocuous picture that we painted in chapter 5 when we described the temporal horizons of prolongation, an operation whose ontological ambition seemed quite unpretentious. By placing the problem of property at the heart

THE CARE OF THINGS

of public debate, conflicts around the right to repair reveal the political aspect of each maintenance operation, which still raises the question of the relationship that humans have with things, well beyond the neutral vocabulary of use and consumption.

Over these debates looms the problem of autonomy. This is clear in the case of farmers deprived of the room for manoeuvre that had until then been left to them by agricultural machines of which they had learned to handle the components, adjust the settings or replace the parts. The issue is fundamental for all the groups that have defended the right to repair since these questions first arose and, in the United States at least, it has long supplanted environmentalist arguments. According to these collectives, the purchase and the act of consumption must be considered as operations of definitive detachment, which free the buyers from any link with the manufacturers and authorize them to act with the objects as they see fit, without manufacturers having any say. Any attempt to obstruct repair is from this point of view a symptom of unacceptable control by one party over another.

Apart from the matter of relations with industrialists, there are among the defenders of the right to repair traces of an even more general desire for emancipation, one that is reminiscent of the ideals of the anti-consumerist movements of the 1970s. Countering the emancipation offered by consumerism, and its promise of liberating and carefree uses, activists propose a higher level of emancipation: one that liberates consumers from the yoke of contemporary technology. It is the influence of technical objects themselves that must be thwarted, as these objects have an unfortunate tendency to close in on themselves without leaving space for their owners.[45] Like sociologists and anthropologists of technology, repair activists strive to open the 'black boxes' that a large number of mass consumer products now consist of. Telephones and computers have been the paradigmatic figures of this type of object for several years. The activists' gesture of emancipation here involves an art of dismantling in which iFixit teams became masters, publishing over the years detailed reports of the systematic disassembly that they subject each new Apple product to.

This desire not to bow to the passive figure of the disciplined user is found even in Repair Cafés. The latter have been flourishing almost everywhere in rich countries for several years, and their spokespersons highlight, in addition to the environmental virtues of extending the lifespan of objects, an emancipatory aim in line with the writings of Richard Sennett and Matthew Crawford.[46] These initiatives are also directly linked to the American counterculture of the 1960s, including its libertarian side.[47] Being able to repair the objects you possess also means being able to remain free and independent whatever happens.

If we follow this last line of argument, we can conclude that the emancipation claimed by the defenders of the right to repair distances them from the oscillations of attention described in the previous chapters, which testify to a common becoming of things and humans. And it is true that, among some of the activists in question, one can sense a tendency to spotlight strong men and women, whose ability not to let themselves be 'alienated' in an asymmetrical relationship to recalcitrant technical objects is lauded. But to stop at this very Cartesian idea of power would be premature. Because the things from which humans have to 'free themselves' through their ability to disassemble them and extend their duration are not 'all things', but very specific mass consumer goods, deliberately locked by industrialists. If we take this idea to its logical conclusion and return to the vocabulary that we have adopted throughout this book, we can even conclude that the demands for the right to repair amount to defending the possibility of taking fixed *objects* with material boundaries made artificially watertight, and treating them as open *things*, still in the making. These demands are a request for duration: they literally plead the cause of the thing to be maintained. The power that is to be gained here is not part of a balance of power with the things themselves. It lies in the refusal of absolute control that is put forward, with considerable technological and legal support, by industrialists. It is the possibility of a dance of maintenance that is at stake – the ability to maintain a caring relationship with things that we not only use, but we also get to know.

Redistributed knowledge

The resistance of defenders of the right to repair argues for disassembly and against the tendency of modern objects to remain black boxes. It is also an injunction to leave open spaces for inquiry and to cultivate relationships with things that promote a process of learning rather than passive consumption. As Steward Brand claims in his book on building maintenance, 'maintaining is learning'.[48] Defending the right to take care of things therefore not only means renegotiating the ownership of consumer objects and fighting for the autonomy of users, but also simultaneously claims a right to knowledge.

This question lay at the centre of the very first struggles for the right to repair. It was around diagnostic practices that conflicts between independent repairers and automobile manufacturers crystallized in the United States.[49] Following the implementation of the Clean Air Act, manufacturers were required to equip their vehicles with standardized devices aimed at measuring their emissions. It was in this context that the integrated diagnostic tool OBD-II (for On-Board Diagnostic) was created: all cars were equipped with it from 1996 onwards. The instrument is fitted with a connector to which a control interface giving access to diagnostic information can be linked. Just as the implementation of elements of software code, coupled with the constraints of the Digital Millennium Copyright Act, has been mobilized by many manufacturers to regain control over the destiny of certain objects, the OBD-II and its connector have been used as barriers to the knowledge of vehicles. By maintaining control over the interfaces, manufacturers had a simple and effective way to permanently exclude independent repairers from the circuit and verticalize the automobile maintenance sector. Faced with the danger, professional associations (the Automotive Aftermarket Industry Association, now the Auto Care Association, and the Coalition for Auto Repair Equality) got together to demand full access to diagnostic tools and repair instruments for all independent garages, thus laying the foundation stone of the 'Right to Repair' movement. The legal battle lasted several years, and it was only in 2011 that a first law was passed by Congress: the Motor Vehicle Owners' Right to Repair Act, which requires, among other things, that automobile manufacturers 'provide to the vehicle owner and service providers all information necessary to diagnose, service, maintain, or repair the vehicle'.[50]

The role of knowledge in the care of things extends well beyond the boundaries of the right to repair movement. During her investigation in Uganda, Lara Houston highlighted the crucial place of the circulation of knowledge in repair workshops.[51] In these places, which can seem crude and even anarchic in the eyes of a Western observer, local forms of knowledge preciously cultivated by repairers from the same neighbourhood are exchanged. This knowledge takes the form of gestures, tips and stories that perpetuate singular experiences in narrative form and aggregate them around a collective. At the same time, other types of knowledge circulate internationally in these repair workshops, which always have a computer connected to online platforms. Just as farmers sought pirated software from Polish and Ukrainian hackers, repairers in Kampala spend their time consulting forums through which they connect with many other repairers scattered around the world. These forums bring together a sometimes competitive community in which very different people rub shoulders (technicians, hackers, passionate amateurs…). It is through these intermediaries that repairers can find solutions to specific problems, order tools to 'flash' old phones to be repaired, or obtain manuals and other technical plans.

Nicolas Nova and Anaïs Bloch also insist on the importance of the exchange and production of knowledge associated with the practices of maintenance at play in telephone repair stores in Switzerland, a context which does not differ as much as one might initially imagine from that of Uganda.[52] They identify three main dimensions. Repairers acquire situated and incorporated knowledge over time, which allows interventions to take place in the best possible conditions, drawing alternately on gestures that have become routine and improvisations. This largely tacit knowledge is complemented by the ingenuity of the reverse engineering practices used by repairers. In the vein of the demonstrations given by the iFixit teams, they seek not only to identify the right ways to repair phones, but also to understand how they work – both being essential. Finally, here too, local and international cooperation networks, woven from face-to-face meetings and online exchanges, are crucial. They create full-fledged circuits of 'globalization from below'.[53] To highlight the richness of this knowledge, which pervades repair workshops (often denigrated and relegated to the background of the information society), Nova and Bloch compare them to laboratories of reverse research and development – 'laboratories of care',[54] within which an art of caring DIY is cultivated, and which are ultimately just as effective as the much more presentable and widely publicized projects of 'makerspaces' and other 'fablabs'.

The struggles of right-to-repair activists, as well as the independent practices of the myriad repairers who operate on the periphery of official circuits, show that the question 'who takes care of things?' is directly linked to that of the circulation of knowledge. This affects instrumental knowledge, of course (how to properly maintain this or that object, how to replace a part, etc.) – but not only that. As we saw in chapters 3 and 4, maintenance activities facilitate the development of knowledge about the things themselves, by discovering dimensions and behaviours that cannot be grasped in 'normal' usage. The particular moment of maintenance, like that of restoration, offers opportunities for the thing to ensure it 'makes itself knowable and lets itself be known'.[55]

If the promoters of the right to repair also fight for a right to knowledge, this is not only in the name of an access that needs to be negotiated so that a greater number of people can participate in the care of things. They also aim to ensure the best possible conditions in order to generate and cultivate specific knowledge, as close as possible to the material and its transformations – a knowledge too often discredited, or even deliberately blocked, in maintenance configurations controlled by manufacturers.

The people of things

It is necessary to underline one last fundamental aspect of the demands of the protagonists of the right to repair, one that we started to sense when discovering the importance of knowledge in their struggles. Alongside individual autonomy and environmental issues, the different collectives all insist on the value of the maintenance and repair occupations. Far from being confined to the libertarian idea of the independent user on whom no big industrial group can impose its law, the campaigns in favour of the right to repair mobilize the entire field of activity of the after-sales jobs, which represent a very important source of employment. On the French version of the iFixit website, for example, one comparison shows that a thousand tons of electronic devices generate two hundred jobs if they are repaired, only fifteen if they are recycled, and barely a single one if they are taken to the landfill. In its latest report devoted to repair, the French Environment and Energy Management Agency estimated that in 2018 the sector had 125,000 companies in France, accounting for 226,000 jobs (including around 152,000 salaried jobs), and that it generated a total turnover of 26 billion euros.[56]

The argument is all the more well founded as it concerns a network of local jobs throughout all the territories of a country. It even took a new turn during the pandemic, which saw supply chains blocked and spare-parts circuits dry up. As early as 5 March 2020, even before embarking on the development of a database of medical-equipment repair manuals, Wiens explained that a strong repair sector represents a considerable asset in dealing with these radical disruptions to the globalized economy. In his view, an ambitious right to repair, which gets rid of the artificial constraints of intellectual property and allows a dense network of independent repairers to develop, constitutes an essential element of national 'resilience'.[57]

Apart from providing an economic demonstration, this argument comprises a political operation by describing a world where mainte-nance and those who practise it matter. It redraws the landscape of mass consumption by encouraging us to place in the foreground the crowd of people associated, in one way or another, with the existence of consumer objects usually presented as entities crystallized in the impassive finitude of their form and their functions – the same crowd of women and men that Mierle Laderman Ukeles has never ceased to laud in her artistic performances, and who live in the interstices of the overly neat categories of 'design' and 'use'. Highlighting this economic and occupational world of maintenance and repair, defenders of the right to repair 'repopulate' the disembodied descriptions of contemporary material life, indirectly responding to the sociologists who have similarly repopulated the descriptions of the social world with the 'missing mass' of the artefacts making up that world:[58] they remind us that nothing, from everyday consumer goods to gigantic infrastructures, exists or lasts without a few human beings attached to it, well beyond the framework of 'consumption' alone. From this point of view, their plea echoes the early work of Madeleine Akrich, which showed how the absence of taking these humans into account could lead to obvious failures in the trajectory of technical objects, such as the photovoltaic kit sent from mainland France to French Polynesia in the 1980s and whose very design prevented its potential users from resorting to local repairers in order to adjust its settings and repair it in the event of a breakdown.[59]

At a time of high-tech capitalism which features 'bright and shiny' objects[60] and increasingly relies on promises of technological autonomy, the question of the place given to this 'people of things', to the knowledge that they cultivate and the deeds they perform in the name of the modest temporal horizon of prolongation, is crucial. As Nova and Bloch point out,[61] the world of independent repair, partly unofficial and even illicit, operates on the model of the 'patches' that Anna Tsing describes: those sites where economic value is produced and trade is generated outside of the rationalized and controlled circuits of globalized capitalism, trade that is essential to its overall functioning but remains confined to its margins.[62] And if the promoters of the right to repair insist on the territorial roots of these 'patches', their argument has an obvious geopolitical significance. We mentioned this attentional shift in chapter 3: a large proportion of this people of things lives and works far from the eyes of consumers quite ignorant of it, in Africa or Asia, in countries where *objects* with accelerated obsolescence become *things* again, things that can last. Telephones, computers, clothing, cars and household appliances are disassembled, reconditioned, tinkered with, and have their lives extended, very often at the cost of painful, toxic manipulations, erased from the triumphant history of the objects of mass consumption from the countries of the North.

Responsibilities

By tracing the main conflicts that were organized around the question of the right to repair, we have shown how political is the question 'who takes care of things?' Beyond the problem of a 'dirty work' that it would be enough merely to recognize, maintenance is also an operator of autonomy, an activity through which power is redistributed in the face of consumer goods, and especially in the face of the firms that produce them. And, as the case with which we started this chapter illustrates, maintenance is also sometimes a matter of responsibility. When Wiens and his team decided to tackle the problem of the maintainability of ventilators in the midst of the Covid-19 crisis, they were getting involved in a task that was close to their hearts. They were committed. Beyond the struggles between activists demanding the right to repair and industrialists wishing to keep control of objects in order to control their destiny, the resistance at work in certain kinds of maintenance thus takes the form of an act of volunteering, of assuming responsibility for what is deemed to be the imperative need to take care of certain things. The question of the division of labour then takes another angle, expressed perfectly in the lapidary expression 'Who cares?' Who cares about this thing or that? The responsibility expressed here sends us to the definition that Haraway gives, one that plays on the composition of the word itself: 'response-ability'.[63] From this perspective, 'care is not only a practice and an action turned outwards, but also a *desire to respond*'.[64] Taking responsibility means both responding to the call of fragile things and being responsible for their becoming, making oneself obligated towards them.

To illustrate this point and close this chapter, let's return to the charismatic participants of Untergunther. It would be an exaggeration, of course, to equate their actions with the right-to-repair movement. That said, it is important to take stock of its political significance. Clandestine restorers act as maintenance activists. We can guess as much from their name, which the press presented as a tribute to hypothetical dogs whose recorded barking made it possible to scare away nosy people who got too close to construction sites, but we can also assume it is a thinly veiled reference to the philosopher Günther Anders. When we were able to ask him by phone, Untergunther spokesperson Lazar Kunstmann laughed and said he didn't know. Fair enough – but the connection doesn't seem completely far-fetched. Anders was a student of Husserl and close to Heidegger, and an eminent critic of modernity and the race for technological supremacy; he was concerned about the imminent disappearance of humanity, a concern that we find in Kunstmann's writings and recordings, notably during his speech at the Long Now Foundation.[65] The restoration operations carried out by the collective are presented as political and moral actions – as the response to an imperative which is no longer endorsed by anyone: the care of a heritage in decline. Untergunther's mission, Kunstmann explained, consists of coming to the aid of the 'urban neglected', the countless sites which bear witness to the most beautiful things humanity has done, but about which neither the State nor municipalities care any more.

'For Soufflot and against carelessness', we can read in the excerpt of the work he devoted to the case of the Panthéon clock.[66] It would be difficult to make it any clearer. The challenge of the restorations performed by the collective lies in remedying a situation of abandonment, literally an absence of care. The aim is to make the few things for which they can act last, hoping that a 'less superficial society' will eventually emerge, one that will once again be able to make them objects of attention.[67] We find many barbed comments in Kunstmann's various speeches: he despairs over the inaction of the administration and the absence of any sense of history on the part of its representatives. And the episode of the Panthéon clock is significant from this point of view, since it means we discover that an object, grand as it may be, forming part of one of the most prominent monuments in Paris, and located at the heart of the city, had been abandoned for several decades. This isn't a total surprise, obviously. It would be naïve to imagine that each building, each artefact presenting a certain historical value, could be the subject of special attention, maintained and preserved in the best possible conditions. But the example is no less revelatory. The clock seemed to have disappeared from the concerns of those responsible for the Panthéon and of the Centre for National Monuments, quite a powerful institution. According to Kunstmann's account, the clock had become literally invisible; the site's administrator seemed to become aware of its existence thanks to the revelation of its clandestine restoration, which gave the Untergunther team the opportunity to take responsibility, for a while, for the clock's becoming.

During the telephone interview he gave us, Kunstmann clarified that this gesture of taking responsibility was closely linked to the proximity that the members of the collective cultivated with the different things they took care of, which they encountered almost daily. Their attachment to the catacombs and certain places in Paris, first and foremost the Panthéon, had made them 'obligated' to a neglected heritage which they respected and which they did not want to see disappear. This experience, which reminds us that the ethics of maintenance is always *situated*, in touch with a specific environment, is accessible to everyone, says Kunstmann. You just have to look around and take the trouble to pay attention to the things that call for maintenance. You just have to take responsibility, in the sense of being 'affected' by these things. Of course, there is nothing obvious in this gesture, as evidenced by the crowd of catacomb enthusiasts who are not at all concerned with preservation, or who prefer to preserve other types of objects, such as the regularly restored wall frescoes.[68]

The concerns of the promoters of the right to repair obviously differ from the concerns of Kunstmann and his comrades. This is primarily due to the types of things that they are attached to. If the heritage which worries the members of Untergunther is neglected, it is nevertheless much more 'noble' than the consumer goods over which consumer associations, independent repairers and large manufacturers argue. And yet the two approaches come together in the defence of a caring and responsible position that consists of acting knowingly, getting involved in the sometimes difficult dance of the entangled becoming of things and the humans who accompany them. In this respect they agree with Mierle Laderman Ukeles, for whom, notes Caroline Ibos, 'maintenance is not a peripheral and contemptible question, not just a dirty work, but a collective responsibility as well as a political question of primary importance'.[69]

Precisely because they insist on the possibility of simply extending the life of innocuous objects, the many people of things who defend, in one way or another, a right to repair continue to keep this political question alive. By paying attention to these resistances and conflicts, we realize that we must constantly open up the problem of the distribution of care activities to the things themselves, but also to the less obvious partners in the dance of maintenance. We must split the question into two even if (or rather because) the first one can't exist without the second one: who takes care … of what?

Conclusion

The ambulatory approach favoured throughout these pages – which has allowed us to walk across extremely diverse sites, to come across a multitude of apparently unconnected people, and to discover, through their hands and their eyes, countless things with heterogeneous contours and destinies – aimed to highlight the ethical and political significance of maintenance activities while remaining sensitive to their irreducible variety. These wanderings, as we observed and listened to those who take care of things, have opened several avenues that we intend to retrace before coming to our journey's end.

First, we realized that, through numerous forms of maintenance, an alternative path was being invented in the interstices of a capitalist regime massively organized around an ever-growing dynamic of consumption and production. In a world where replacement and its corollaries, the depletion of resources and the generation of waste, have become the driving forces of the economy, making things last is an almost subversive operation. Even though it obviously does not represent the only solution to the damage of overconsumption, maintenance appears as one of the forms of action by which humans can create *a slightly more* habitable world, by partly short-circuiting the logics which contribute to its growing uninhabitability.

This short-circuiting takes place in the very particular temporalities that maintenance activities cultivate. The care of things is a daily pulse, a ritornello that never stops. It weaves from the daily life of modernity timelines that escape both the blind headlong dash of progress and the prophesied disaster.[1] Maintenance unfolds the present and thickens it with a duration that is always in the making. It opens up possibilities, not through any break with the past or projection into the future, but in the practical invention of continuity.

To fully understand this, however, we must get rid of a stubborn reflex that associates maintenance with the politics of the status quo. The detailed descriptions of the gestures of the maintainers have taught us that, even when it is part of a preservationist programme, even when its horizon is permanence, the care of things is always generative. Whatever its pretensions, it cannot escape the perpetual transformations of matter. Far from preventing these transformations or circumventing them, the women and men who practise maintenance contribute fully to them. Maintenance is always a form of creation, part of a common becoming, and never a simple reproduction. Thus, if it is political, it is not because it diverts human beings from progress and disaster, and thus extricates them from the flow of time, but on the contrary because it can force them to invent adequate ways of participating in it.

As a focus of critical activities, maintenance is also a diplomatic site where relationships are forged between humans and things – relationships that go far beyond the registers of the instrumental and the symbolic. By making material fragility a starting point, a common condition to which it is crucial to become sensitive, maintenance deploys an 'immanent attention' which is the organizing principle of these relationships.[2] It requires tact, not only by adapting one's gestures and use of tools to each situation, but also by providing space for the things themselves. The attentional skills at work in maintenance aim to bring to the fore the incessant material transformations which are neglected, or even hidden, in most situations of the consumption and use of modern artefacts. This material diplomacy refuses to play along with a world where technology is always solid and consumer goods are locked in just two possible states: indefinitely shiny and new, or doomed to obsolescence. The sensible encounters on which the care of things is based require us to allow them the possibility of manifesting themselves.

In the preface he wrote for the French edition of *How Forests Think* by Eduardo Kohn, Philippe Descola remarks that the semiotic generalization carried out by its author stops at the borders of a controversial definition of the living world.[3] Indeed, if animals, but also forests and rivers, can think and signal in Kohn's writings, this is not the case for stones, for example. Descola regrets this limitation, which leaves out too many non-human beings in his eyes. All things considered, maintenance can teach us to go beyond this limitation, and encourage us to take into consideration the signs generated by many of what Descola calls 'biotic' beings: stones, and the multitude of composite objects that give shape to the human world.[4] Kohn was inspired by the Runa of Ávila, in particular their understanding of dreams, to grasp the capacity that many organisms have to signal to each other, beyond or below the symbolic relationship, which anthropology is generally obsessed with. Our source of inspiration is far closer, and its interest lies precisely in the fact that it is completely integrated into the fabric of modernity, unlike the populations observed by anthropologists keen to make an ontological turn by departing from the 'self-evident' naturalist cosmology of the West and the North. By observing the women and men who practise maintenance, by paying attention to their own sensitivity to the fragility and permanent mutations of matter, we discover that it is possible to treat the most innocuous *objects* as incomplete *things*, capable of multiple forms of expression.

But a question haunts every diplomatic initiative: what is the balance of forces that pre-exist it? If it is valuable to highlight poorly understood forms of negotiation and to emphasize the importance of tact in taking care of things, it is also important to ask who, and what, has the right to participate in negotiations. This third aspect of the politics of maintenance is where our own description is grafted onto that of the maintainers. As we have stated on several occasions, our objective in this book was also to make maintenance and those who practise it *matter*. Making maintenance matter means showing from a new angle the ways in which humans use artefacts of all kinds to inhabit the world. It means showing that if these objects persist, if they can act and participate in the fabric of human societies, it is not only because they were designed and manufactured in particular conditions, it is also because women and men take care of them and never stop considering them as things in the making. But these women and men are left in the background of the triumphalist stories of innovation and technical progress, particularly when maintenance is their job. It is therefore a gesture of repopulation that we sought to carry out and which we call on others to continue, in line with the conceptual art of Mierle Laderman Ukeles and the sociology of Susan Leigh Star. In addition to highlighting the crucial role that objects play in the constitution of human societies, this 'missing mass'[5] that the social sciences have long neglected, the study of maintenance calls for taking into account the multitude of people – hidden hands or recognized experts – who ensure things can exist and last. A

CONCLUSION

very heterogeneous 'people of things' who work on a daily basis to ensure the continuity of the sociomaterial fabric of the world.

The challenge here is thus to recognize the inseparable links that unite things to women, men and their environments in a common becoming. The expression 'care of things', an expression that we value, is a way of emphasizing this political aspect of attention to maintenance, explicitly associated with feminist theorists of care, who have ceaselessly fought against the myth of an 'autonomous' human economy whose value is gauged solely on the permanent concealment of reproductive labour and the interdependencies that bind humans together.[6] Placing maintenance at the forefront means refusing another myth: that of the autonomy of the technical and artefactual part of the world. The concern for things is always also a concern for the women and men who take care of them. It encourages us to give a name and a face to the people who are responsible for making them last. But far from being limited to an issue of visibility, it forces us to question the place attributed to the maintainers' own expertise, to their ways of getting to know things. It also forces us to question the conditions in which these people work. What is the human cost of making things last? How are the bodies of those who take care of this or that artefact affected? What consequences does maintenance work have on their health? And furthermore, how is this work valued? What part does it play in the chain of values of contemporary capitalism?

On this point, we must also recognize that we have not encountered many women in our stories, like in Ukeles' (except herself). As if caring for things were reserved for men, while caring for people would be reserved for women. However, we know that this is not the case, and that, particularly in the domestic sphere, a huge set of material care activities are carried out by women.[7] This domestic work – the gestures that are deployed in it, its specific concern for continuity, and the very particular things to which it relates – is one of the lines of flight of our inevitably unfinished exploration.[8]

As Joan Tronto notes with regard to care in general, if the question of the people concerned is important, it is also because it highlights the profound inequalities which separate those who have the luxury of not worrying about the fragility of the world from those who cannot help but be sensitive to it and work to minimize its consequences.[9] In the case of maintenance, this means that knowing who takes care of

THE CARE OF THINGS

things goes hand in hand with another question: who can afford not to take care of them? If it is not possible or desirable to imagine that each human being should constantly be concerned about the fragility of all the things around them and take an active part in their maintenance, we can still see, in the recognition of this fragility and the incessant work that accompanies it, a path to better distribute some of these tasks and to make the ethics of care a critical engine of a technical democracy which does not stop at the borders of innovation and design – a path that creates ever more 'sensitive experiences likely to affect us, to transform our ways of feeling the world and acting in it politically'.[10]

Finally, the last facet of the politics of maintenance touches on the complex question of the value given to the things at the centre of care. This question reminds us that not all maintenance is good in itself and that, if we have sought to highlight what maintainers could teach us about the world, we have in no way formulated a general argument in favour of maintenance. To imagine that one could be 'for' or 'against' maintenance in general would be to ignore the variety of concrete situations in which maintenance is carried out. Above all, it would forget the highly problematic question of identifying the things that need to be taken care of. How can we choose, among the multitude of objects that make up the human world, those whose duration we must cultivate? What should we take care of? In what capacity, by what right? The question is all the more delicate because it obviously comes with another: what does not deserve, or no longer deserves, to last?

This lies at the centre of the 'disconnectionist' ecology recently defended by Emmanuel Bonnet, Diego Landivar and Alexandre Monnin,[11] who call for the organization of dismantling operations in order to literally interrupt the deleterious effects of globalized financial capitalism and its infrastructures which colonize everything in their path while being subject to maintenance, however expensive. Working towards closure, for what they call 'destoration', is an essential gesture whose great merit is to refuse to brush under the carpet the question of artefacts and, more generally, of the 'technosphere', a crucial question in the debates entailed by the environmental crisis, and one that contemporary calls to take care of the living world too often neglect. This invitation to think about closure takes a very concrete practical form, which involves in particular the collective production of cartographies aimed at identifying and discussing the web of interdependencies and thus the threads to be 'cut' in order to interrupt the existence of this or that technology. It is a matter of bringing into politics the difficult question of the technical heritage for which one sector of humanity bears responsibility.

Maintenance represents, in a way, the other side of this political issue, its 'positive' side. If it is important to emphasize the need to organize ourselves if we are to interrupt the life of certain objects and certain infrastructures, this must not suggest that it is enough to allow other artefacts, those that we agree to designate as just and desirable, simply to continue to exist as if this continuity were self-evident. Likewise, certain materials and certain infrastructures require that we still work to maintain them, either because we do not yet have solutions allowing them to be closed down definitively, or because they are stubborn and resistant to disintegration.[12] Making things last requires that we invest in their care in one way or another. A politics of maintenance involves not only the description of these investments and these commitments (physical, material, reflexive, financial, etc.), but also a debate as to their nature – especially since maintenance is by no means a transparent operation. Always transformative, its political impact is as much due to the choice of things on which it acts as to the ontological enactments each intervention performs. Maintaining always means making certain aspects of things matter more than others.

What is maintained (what things and what dimensions of the thing)? What is not? These are two sides of the same essential question if maintenance is to play a part in composing a more habitable world. The answers to this question cannot be based on values calculated in advance that would make it possible to produce a priori hierarchies. There is a great risk, in fact, of making these problems a matter of priorities and of reducing the politics of maintenance to the establishment of a list of things 'that matter'. This individualist logic would amount to adopting what Catherine Larrère calls, in the field of ecology, a 'biocentric' logic,[13] which consists of always increasing the number of entities that must be protected, without taking into consideration their interdependencies. The authors of the 'new ecology' have shown this clearly: if we want to create a habitable world, we must not so much increase the host of beings to be taken into consideration as concern ourselves with the community as a whole and the links that constitute it.

> Contrary to the object-ontology of classical physics and biology in which it was possible to conceive of an entity in isolation from its milieu – hanging alone in the void or catalogued in a specimen museum – the conception of one thing in the New Physics and New Ecology necessarily involves the conception of others and so on, until the entire system is, in principle, implicated.[14]

This 'ecocentric' ethic[15] can inspire an ambitious and desirable politics of maintenance. It encourages us to develop forms of care for things that are aware of the multiplicity of material interdependencies and cultivate attention to the fragility of milieus and the relationships that are woven there – a maintenance which assumes that making something last also means continuing to bring into existence 'the world that comes with it'.[16]

From this point of view, maintenance involves acting without an abstract definition of the things that need to be taken care of. It requires fully accepting the fact that the material diplomacy of care is deployed *in a situation*, through open, always generative, inquiries. Seen in these terms, maintenance is less a matter of values already there from which one needs to choose, and more an open and uncertain process of *valuation*,[17] through which values in the broad sense are progressively identified and discussed. This perspective is in line with William Cronon's critique of the idea of wilderness[18] in the world of environmental conservation. The problem with this notion is precisely that it produces dogmatic hierarchies and ends up isolating supposedly wild and pristine parts of the world from the rest, in the name of their preservation. Wilderness depoliticizes the act of conservation, even as it produces irreversible inequalities. Since the choice of what care should be focused on is always double-edged, wilderness, like any value fixed in advance, generates, at the same time as it designates a list of sites to be protected, a considerable quantity of territories not worthy of preservation. Thus, '[m]y principal objection to wilderness is that it may teach us to be dismissive or even contemptuous of such humble places and experiences'.[19] But can we go so far as to completely dismiss the call of the preservationists? Rather than getting rid of the idea of the 'wild', Cronon suggests broadening its scope and generalizing it. He encourages us to be attentive to the wild wherever it is, to the things that are worth maintaining and protecting, including in our own garden, however unnatural it may be, at the centre of which stands a tree that perhaps also deserves our attention and care.

CONCLUSION

It is by carrying out inquiries, and going so far as to take into consideration the 'cracks of a Manhattan sidewalk',[20] that humans can discover how to assume their responsibilities and better play their role on Earth.

This invitation echoes the approach that we have adopted in this book by affirming that taking into consideration the fragility of things, in the same way as that of people and environments, and recognizing the value of their maintenance, helps us to compose adjusted ways of inhabiting the world without ever escaping from it. For this, it is valuable to learn, simultaneously, to become sensitive to things, including cracked pavements, and to debate what attaches us to them, while paying attention to the activities, concerns and knowledge of those who take care of things on a daily basis.

Notes

Introduction

1 See, among many other studies: Léna Balaud and Antoine Chopot, *Nous ne sommes pas seuls. Politique des soulèvements terrestres* (Paris: Seuil, 2021); Emanuele Coccia, *Sensible Life: A Micro-ontology of the Image*, translated by Scott Alan Stuart (New York: Fordham University Press, 2016); Gilbert Cochet and Béatrice Kremer-Cochet, *L'Europe réensauvagée. Vers un nouveau monde* (Arles: Actes Sud, 2020); Vinciane Despret, *Habiter en oiseau* (Arles: Actes Sud, 2019); Dusan Kazic, *Quand les plantes n'en font qu'à leur tête. Concevoir un monde sans production ni économie* (Paris: La Découverte/Les Empêcheurs de penser en rond, 2022); Baptiste Morizot, *On the Animal Trail*, translated by Andrew Brown (Cambridge: Polity, 2012); Estelle Zhong Mengual, *Apprendre à voir. Le point de vue du vivant* (Arles: Actes Sud, 2021).

2 Baptiste Morizot, *Ways of Being Alive*, translated by Andrew Brown (Cambridge: Polity, 2022), p. 1.

3 Emmanuel Bonnet, Diego Landivar and Alexandre Monnin, *Héritage et fermeture. Une écologie du démantèlement* (Paris: Divergences, 2021). For an overview of recent dynamics of detachment that operate in different domains, see Frédéric Goulet and Dominique Vinck (eds.), *New Horizons for Innovation Studies. Doing Without, Doing with Less: Destabilisation, Discontinuation and Decline as Horizons for Transformation* (Northampton, MA: Edward Elgar, 2023).

4 Bruno Latour, *La Clef de Berlin. Petites leçons de sociologie des sciences* (Paris: La Découverte, 1993).

5 Bruno Latour, *Facing Gaia: Eight Lectures on the New Climatic Regime*, translated by Catherine Porter (Cambridge: Polity, 2017).

6 There are countless high-calibre works that consider the transformation of living things and the modes of presence of the products of human activity, from major infrastructures to residual products, as forms of toxicity, and more generally of the deterioration of the environment. See, among many others, Bernadette Bensaude-Vincent, 'Plastics, materials and dreams of dematerialization', in Jennifer Gabrys, Gay Hawkins and Mike Michael (eds.), *Accumulation: The Material Politics of Plastic* (Abingdon: Routledge, 2013),

NOTES TO PP. 8–9

pp. 17–29; Sophie Houdart, 'Les répertoires subtils d'un terrain contaminé', *Techniques & Culture*, no. 68, 2017, pp. 88–103; Max Liboiron, Manuel Tironi and Nera Calvillo, 'Toxic politics: Acting in a permanently polluted world', *Social Studies of Science*, vol. 48, no. 3, 2018, pp. 331–49; Matthieu Duperrex, *Voyages en sol incertain. Enquêtes dans les deltas du Rhône et du Mississippi* (Marseilles: Wildproject, 2019); Soraya Boudia, Angela N. H. Creager, Scott Frickel, Emmanuel Henry, Nathalie Jas, Carsten Reinhardt and Jody A. Roberts, *Residues: Thinking through Chemical Environments* (New Brunswick, NJ: Rutgers University Press, 2021); Trisia Farrelly, Sy Taffel and Ian Shaw, *Plastic Legacies: Pollution, Persistence, and Politics* (Athabasca, Alberta: Athabasca University Press, 2021).

7 Hilary Sample, *Maintenance Architecture* (Cambridge, MA: The MIT Press, 2018), p. 17.

8 Yves Citton, *The Ecology of Attention*, translated by Barnaby Norman (Cambridge: Polity, 2017), p. 161.

9 Jérôme Denis and David Pontille, *Petite sociologie de la signalétique. Les coulisses des panneaux du métro* (Paris: Presses des Mines, 2010).

10 Jérôme Denis and David Pontille, 'Maintenance epistemology and public order: Removing graffiti in Paris', *Social Studies of Science*, vol. 51, no. 2, 2021, pp. 233–58.

11 Daniel Florentin and Jérôme Denis, *Gestion patrimoniale des réseaux d'eau et d'assainissement en France*, Caisse des Dépôts-Institut pour la recherche et Banque des territoires, 2019.

12 For an overview, see Jérôme Denis and David Pontille, 'Why do maintenance and repair matter?', in Anders Blok, Ignacio Farías and Celia Roberts (eds.), *The Routledge Companion to Actor-Network Theory* (Abingdon: Routledge, 2020), pp. 283–93.

13 Since the first publication of this book in French, a large number of publications have enriched the flourishing domain of 'Maintenance and Repair Studies'. Among them, we would like to mention a few important collective books, which we encourage readers to discover: Markus Berger and Kate Irvin, *Repair: Sustainable Design Futures* (London and New York: Taylor & Francis, 2022); Stefan Krebs and Heike Weber (eds.), *The Persistence of Technology: Histories of Repair, Reuse and Disposal* (Bielefeld: Transcript Verlag, 2021); Dimitris Papadopoulos, Maria Puig de la Bellacasa and Maddalena Tacchetti (eds.), *Ecological Reparation: Repair, Remediation and Resurgence in Social and Environmental Conflict* (Bristol: Bristol University Press, 2023); and Mark Thomas Young and Mark Coeckelbergh (eds.), *Maintenance and Philosophy of Technology: Keeping Things Going* (London and New York: Taylor & Francis, 2024).

NOTES TO PP. 9–15

14 Narratives are a resource for producing knowledge about the world that the formal procedures of the modern sciences do not capture. See Isabelle Stengers and Vinciane Despret, *Women Who Make a Fuss: The Unfaithful Daughters of Virginia Woolf*, translated by April Knutson (Minneapolis, MN: University of Minnesota Press, 2021).

15 Like the 'sensitizing concepts' dear to Herbert Blumer, who says: 'Whereas definitive concepts provide prescriptions of what to see, sensitizing concepts merely suggest directions along which to look' (Herbert Blumer, 'What is wrong with social theory?', *American Sociological Review*, vol. 19, no. 1, 1954, p. 7).

16 Dorothy E. Smith, *Institutional Ethnography: A Sociology for People* (Lanham, MD: Rowman & Littlefield, 2005).

17 Bruno Latour, *Politics of Nature: How to Bring the Sciences into Democracy*, translated by Catherine Porter (Cambridge, MA: Harvard University Press, 2004); Bruno Latour, 'Why has critique run out of steam? From matters of fact to matters of concern', *Critical Inquiry*, vol. 30, no. 2, 2004, pp. 225–48.

18 Martin Heidegger, *What Is a Thing?*, translated by W. B. Barton Jr and Vera Deutsch (Washington, DC: Gateway Editions, 1968).

19 Latour, 'Why has critique run out of steam?'

20 William James, *The Meaning of Truth: A Sequel to 'Pragmatism'* (1909), available online at: https://www.gutenberg.org/files/5117/5117-h/5117-h.htm. We owe this pragmatist reference to Antoine Hennion: see in particular 'Enquêter sur nos attachements. Comment hériter de William James?', *SociologieS*, 2015, https://doi.org/10.4000/sociologies.4953.

21 Étienne Souriau, *The Different Modes of Existence: Followed by On the Work to Be Made*, translated by Erik Beranek and Tim Howles (Minneapolis, MN: Univocal Publishing, 2015).

22 Tim Ingold, 'Materials against materiality', *Archaeological Dialogues*, vol. 14, no. 1, 2007, pp. 1–16 (p. 9).

23 Karen Barad, 'Posthumanist performativity: Toward an understanding of how matter comes to matter', *Signs*, vol. 28, no. 3, 2003, pp. 801–31.

24 We have arranged these chapters so that they form a course made up of plateaus whose progressive discovery reveals an ever-denser image of maintenance, while leaving space for exceptions and individual cases. That being said, each chapter can be read independently, and nothing prevents you from reading this book in another order and discovering the ethics and politics of maintenance at your own pace. If you want first to understand how collectives defend their way of taking care of things in the face of large industrial groups, you can start with chapter 7. If it is maintenance as a work activity that interests you, and

NOTES TO PP. 15–21

the sometimes rough and ready relationships with things that it involves, you can skip directly to chapter 4. If the controversies and debates on the value of things that need to be maintained and on the role of human beings in their preservation intrigue you, you can turn to chapter 6. If you wonder what changes are brought about by the maintainers' gaze when they scrutinize the place of objects in society, and are curious about the relevance of the vocabulary of care when it comes to talking about things, you will find some answers in chapter 2. If you tell yourself that, above all, the main issue in maintenance is the relationship that humans have with time, you can take a look at chapter 5. If you are fascinated by the ability of maintainers to see signs of wear or failure in objects, chapter 3 may inspire you.

25 It is called *Manifesto for Maintenance Art 1969!*, and is available on the website of the Queens Museum: https://queensmuseum.org/wp-content/uploads/2016/04/Ukeles-Manifesto-for-Maintenance-Art-1969.pdf.

Chapter 1: Maintaining

1 *Manifesto for Maintenance Art 1969!*, available on the website of the Queens Museum: https://queensmuseum.org/wp-content/uploads/2016/04/Ukeles -Manifesto-for-Maintenance-Art-1969.pdf.

2 Andrew L. Russell and Lee Vinsel, 'Hail the maintainers', *Aeon*, 2016, available online at: https://aeon.co/essays/innovation-is-overvalued-maintenance-often -matters-more; 'Let's get excited about maintenance', *New York Times*, 22 July 2017; *The Innovation Delusion: How Our Obsession with the New Has Disrupted the Work that Matters Most* (New York: Currency, 2020).

3 David Edgerton, *The Shock of the Old: Technology and Global History since 1900* (London: Profile Books, 2006).

4 Ibid., p. ix.

5 David Edgerton, 'Creole technologies and global histories: Rethinking how things travel in space and time', *History of Science and Technology Journal*, vol. 1, no. 1, 2017, pp. 75–112 (pp. 76–7), available online at: https://kclpure.kcl.ac .uk/ws/portalfiles/portal/61839321/host_vol1_david_edgerton.pdf.

6 See for example the various special issues devoted to the subject: Lara Houston, Daniela K. Rosner, Steven J. Jackson and Jamie Allen, 'R3pair volume', *Continent*, vol. 6, no. 1, 2017, pp. 1–3; Frederic Joulian, Yann-Philippe Tastevin and Jamie Furniss, 'Réparer le monde. Excès, reste et innovation', *Techniques & Culture*, no. 65–6, 2017, pp. 14–27. Elisabeth Spelman wrote an important essay on this matter at the beginning of the 2000s: see Elisabeth Spelman, *Repair: The Impulse to Restore in a Fragile World* (Boston, MA: Beacon Press, 2002).

NOTES TO PP. 21–25

7 Essential questions are obviously involved in knowing to whom this 'we' actually refers, particularly in connection with the attribution of responsibilities, as evidenced by the numerous debates over the very notion of the Anthropocene and the many alternative terms, from the Capitalocene to the Plantationocene, via the Chthulucene. See Andreas Malm, *Fossil Capital: The Rise of Steam-Power in the British Cotton Industry, c. 1825–1848, and the Roots of Global Warming* (Lund: Lund University Press, 2014); Donna Haraway, *Staying with the Trouble: Making Kin in the Chthulucene* (Durham, NC: Duke University Press, 2016); Janae Davis, Alex A. Moulton, Levi Van Sant and Brian Williams, 'Anthropocene, Capitalocene, … Plantationocene? A manifesto for ecological justice in an age of global crises', *Geography Compass*, vol. 13, no. 5, 2019, available online at: https://par.nsf.gov/servlets/purl/10180270.

8 Spelman, *Repair*.

9 Christopher Henke and Benjamin Sims, *Repairing Infrastructures: The Maintenance of Materiality and Power* (Cambridge, MA: The MIT Press, 2020).

10 Brian Wynne, 'Unruly technology: Practical rules, impractical discourses and public understanding', *Social Studies of Science*, vol. 18, no. 1, 1988, pp. 147–67; Thomas F. Gieryn and A. Figert, 'Ingredients for the theory of science in society: O-rings, ice water, C-clamp, Richard Feynman and the Press', in Susan Cozzens and Thomas Gieryn (eds.), *Theories of Science in Society* (Bloomington, IN: Indiana University Press, 1990), pp. 67–97; Trevor J. Pinch, 'How do we treat technical uncertainty in systems failure? The case of the space shuttle Challenger', in Todd R. La Porte (ed.), *Responding to Large Technical Systems: Control or Anticipation* (Dordrecht: Kluwer, 1991), pp. 137–52. Years later, Bruno Latour drew on the fate of the Challenger shuttle to show how this time the disaster transformed a stable technical *object* into a *thing*, the focus of many concerns. See Bruno Latour, 'Why has critique run out of steam? From matters of fact to matters of concern', *Critical Inquiry*, vol. 30, no. 2, 2004, pp. 225–48.

11 Martin Heidegger, *Being and Time*, translated by Joan Stambaugh, revised edn (New York: SUNY Press, 2010).

12 Jane Bennett, 'The agency of assemblages and the North American blackout', *Public Culture*, vol. 17, no. 3, 2005, pp. 445–65; T. W. Luke, 'Power loss or blackout: The electricity network collapse of August 2003 in North America', in Stephen Graham (ed.), *Disrupted Cities* (New York: Routledge, 2010), pp. 55–68.

13 Geoffrey C. Bowker and Susan Leigh Star, *Sorting Things Out: Classification and Its Consequences* (Cambridge, MA: The MIT Press, 1999).

14 Matthew B. Crawford, *Shop Class as Soulcraft: An Inquiry into the Value of Work* (London: Penguin, 2009).

NOTES TO PP. 26–41

15 The full broadcast is available on YouTube: https://www.youtube.com/watch?v=Wpzvaqypav8.

16 For a discussion and critique of the notion of resilience, see Jeremy Walker and Melinda Cooper, 'Genealogies of resilience: From systems ecology to the political economy of crisis adaptation', *Security Dialogue*, vol. 42, no. 2, 2011, pp. 143–60; Magali Reghezza-Zitt and Samuel Rufat (eds.), *Resilience Imperative: Uncertainty, Risks and Disasters* (London: ISTE Press/Elsevier, 2015).

17 Steven J. Jackson, 'Rethinking repair', in Tarleton Gillespie, Pablo J. Boczkowski and Kirsten A. Foot (eds.), *Media Technologies: Essays on Communication, Materiality, and Society* (Cambridge, MA: The MIT Press, 2014), pp. 221–40.

18 Steven J. Jackson, 'Speed, time, infrastructure: Temporalities of breakdown, maintenance, and repair', in Judy Wajcman and Nigel Dodd (eds.), *The Sociology of Speed: Digital, Organizational, and Social Temporalities* (Oxford: Oxford University Press, 2016), pp. 169–85.

19 Ibid., p. 170.

20 Hilary Sample, *Maintenance Architecture* (Cambridge, MA: The MIT Press, 2018).

21 Several academic and artistic experiments have sought to emphasize the interstitial time of architectural maintenance. The 2015 film by Ignaz Strebel and Susanne Hofer, *Building Care – That's Why Our Cities Do Not Fall Apart*, is a good example of this. It uses plans and drawings to show simultaneously the spaces and the rhythm of interventions by the concierges in charge of the maintenance of a collective building in Switzerland: https://vimeo.com/114462346.

22 The exercise is, of course, valid for other infrastructures. The demand for electricity and its geographical distribution were profoundly transformed in a few days, as were the ways water networks were used. These dramatic changes involved completely new forms of maintenance.

23 Pierre Caye, *Durer* (Paris: Belles Lettres, 2020), p. 64.

24 Spelman, *Repair*, p. 134.

25 Patricia Phillips (ed.), *Mierle Laderman Ukeles: Maintenance Art* (New York: Prestel, 2016).

26 Ibid., p. 101.

27 Ibid., p. 101.

28 Edgerton, *The Shock of the Old*.

Chapter 2: Fragilities

1 Jérôme Denis and David Pontille, *Petite sociologie de la signalétique. Les coulisses des panneaux du métro* (Paris: Presses des Mines, 2010).

NOTES TO PP. 41–50

2 This project, entitled 'Écologies et politiques de l'écrit' ('The ecologies and politics of writing'), was financed by the Agence nationale de la recherche, and was carried out in collaboration with Philippe Artières, Béatrice Fraenkel and Christian Licoppe.

3 Maintenance departments have been a major focus for organization studies, especially in France, even though the latter never problematized the question of maintenance as such, preferring to highlight the question of power struggles between professionals, as in the classic studies by Michel Crozier (see *The Bureaucratic Phenomenon* [London: Tavistock Publications, 1964]), the processes of outsourcing (Annie Thébaud-Monny, *L'Industrie nucléaire. Sous-traitance et servitude* [Paris: Inserm, 2000]), the dynamics of organizational innovation (Gilbert de Terssac and Karine Lalande, *Du train à vapeur au TGV. Sociologie du travail d'organisation* [Paris: PUF, 2002]), and the tensions between planning and the management of the unpredictable that are particularly evident in high-risk sectors (Mathilde Bourrier, *Le Nucléaire à l'épreuve de l'organisation* [Paris: PUF, 1999]).

4 For a more detailed survey of this aspect of metro signs, see Jérôme Denis and David Pontille, 'Maintenance work and the performativity of urban inscriptions: The case of Paris subway signs', *Environment and Planning D*, vol. 32, no. 3, 2014, pp. 404–16.

5 Baptiste Morizot, *On the Animal Trail*, translated by Andrew Brown (Cambridge: Polity, 2012).

6 Ibid., p. 53.

7 Ibid. On this question, Morizot also draws on the texts that Val Plumwood wrote after being attacked by a crocodile. See Val Plumwood, *The Eye of the Crocodile*, edited by Lorraine Shannon (Canberra: ANU Press, 2012).

8 Tim Edensor, 'Entangled agencies, material networks and repair in a building assemblage: The mutable stone of St Ann's Church, Manchester', *Transactions of the Institute of British Geographers*, vol. 36, no. 2, 2011, pp. 238–52.

9 For its English translation, the title became way less provocative: Bruno Latour, 'On interobjectivity', *Mind, Culture, and Activity*, vol. 3, no. 4, 1996, pp. 228–45.

10 Bruno Latour, 'Mixing humans and non-humans together: The sociology of a door-closer', *Social Problems*, vol. 35, no. 3, 1988, pp. 298–310.

11 For a detailed critique of the way the traditional social sciences have grasped the place of objects in society, see in particular the first chapter of Antoine Hennion, *The Passion for Music: A Sociology of Mediation*, translated by Margaret Rigaud (Abingdon: Routledge, 2020).

12 Lucy A. Suchman, *Plans and Situated Actions: The Problem of Human–Machine Communication* (Cambridge: Cambridge University Press, 1987); Jean

NOTES TO PP. 50–54

Lave, *Cognition in Practice: Mind, Mathematics and Culture in Everyday Life* (Cambridge: Cambridge University Press, 1988); Donald A. Norman, 'Cognitive artifacts', in John Millar Carroll (ed.), *Designing Interaction: Psychology at the Human–Computer Interface* (New York: Cambridge University Press, 1991), pp. 17–38; Edwin Hutchins, *Cognition in the Wild* (Cambridge, MA: The MIT Press, 1995).

13 The *Journal of Material Culture*, founded in 1996 on the initiative of Alfred Gell, offers an interdisciplinary forum for fostering work on this theme. In France, the journal *Techniques & Culture* publishes numerous articles on the subject.

14 One of the main criticisms that authors such as Hutchins, Suchman and Lave aim at laboratory psychology is that it artificially isolates the 'mental' processes which it universalizes in its results by emptying the experimental situations of all the artefacts that accompany cognition, perception, computational operations, etc.

15 Madeleine Akrich, 'The de-scription of technical objects', in Wiebe Bijker and John Law (eds.), *Shaping Technology/Building Society: Studies in Sociotechnical Change* (Cambridge, MA: The MIT Press, 1992), pp. 205–24.

16 Carl von Clausewitz, *On War* (various editions, and available online at: https://www.gutenberg.org/files/1946/1946-h/1946-h.htm).

17 This is undoubtedly why so many requests to take the material part of the world into consideration are accompanied by meaningful looks and gestures (banging on the table or lectern, for example) on the part of the person keen to demonstrate its importance.

18 Latour, 'On interobjectivity', p. 236.

19 Latour, 'Mixing humans and non-humans together'.

20 Latour dwells on this point in 'Le groom est en grève', the French version of 'Mixing humans and non-humans together'.

21 Susan Leigh Star, 'Power, technology and the phenomenology of conventions: On being allergic to onions', *The Sociological Review*, vol. 38, no. 1 (supplement), 1991, pp. 26–56; Sociological Review Monograph Series: A Sociology of Monsters: Essays on Power, Technology and Domination, edited by John Law.

22 See especially Tim Ingold, 'Materials against materiality', *Archaeological Dialogues*, vol. 14, no. 1, 2007, pp. 1–16.

23 Tim Ingold, *Making: Anthropology, Archaeology, Art and Architecture* (Abingdon: Routledge, 2013), p. 31

24 Karen Barad, 'Posthumanist performativity: Toward an understanding of how matter comes to matter', *Signs*, vol. 28, no. 3, 2003, pp. 801–31.

NOTES TO PP. 54–58

25 Donna J. Haraway, *Simians, Cyborgs, and Women: The Reinvention of Nature* (New York: Routledge, 1991); Karen Barad, *Meeting the Universe Halfway: Quantum Physics and the Entanglement of Matter and Meaning* (Durham, NC: Duke University Press, 2007).

26 Catherine Larrère, 'Environmental ethics: Respect and responsibility', in Aurélie Choné, Isabelle Hajek and Philippe Hamman (eds.), *Rethinking Nature: Challenging Disciplinary Boundaries* (Abingdon and New York: Routledge, 2017), pp. 54–83; John Baird Callicott, *In Defense of the Land Ethic: Essays in Environmental Philosophy* (Albany, NY: State University of New York Press, 1989).

27 Philippe Descola, *Beyond Nature and Culture*, translated by Janet Lloyd (Chicago, IL: University of Chicago Press, 2014); Eduardo Viveiros de Castro, *Cannibal Metaphysics*, edited and translated by Peter Skafish (Minneapolis, MN: University of Minnesota Press, Univocal Publishing, 2014); Eduardo Kohn, *How Forests Think: Toward an Anthropology beyond the Human* (Berkeley, CA: University of California Press, 2013).

28 Diana Coole and Samantha Frost (eds.), *New Materialisms: Ontology, Agency, and Politics* (Durham, NC: Duke University Press, 2010); Jane Bennett, *Vibrant Matter: A Political Ecology of Things* (Durham, NC: Duke University Press, 2010); Nigel Clark and Bronislaw Szerszynski, *Planetary Social Thought: The Anthropocene Challenge to the Social Sciences* (Cambridge: Polity, 2021).

29 The term 'more than human' is defended in particular by Maria Puig de la Bellacasa, as a way of being able to talk both of 'nonhumans and other than humans such as things, objects, other animals, living beings, organisms, physical forces, spiritual entities, and humans' (Maria Puig de la Bellacasa, *Matters of Care: Speculative Ethics in More Than Human Worlds* [Minneapolis, MN: University of Minnesota Press, 2017], p. 1).

30 Rotor, *Usus/usures. État des lieux/How things stand*, available online at: https://rotordb.org/sites/default/files/2019-06/usus_usures_Rotor.pdf (p. 55).

31 Ibid., p. 17, translation modified.

32 Steven J. Jackson, 'Rethinking repair', in Tarleton Gillespie, Pablo J. Boczkowski and Kirsten A. Foot (eds.), *Media Technologies: Essays on Communication, Materiality, and Society* (Cambridge, MA: The MIT Press, 2014), pp. 221–40.

33 Rotor, *Usus/usures*, p. 57.

34 Bruno Latour, 'Outline of a parliament of things', *Écologie & Politique*, vol. 56, no. 1, 2018, pp. 47–64.

35 Baptiste Morizot, *Wild Diplomacy: Cohabiting with Wolves on a New Ontological Map*, translated by Catherine Porter (New York: SUNY Press, 2022).

36 Rotor, *Usus/usures*, p. 57.

NOTES TO PP. 59–69

37 For the texts in French, see Pascale Molinier, Sandra Laugier and Patricia Paperman (eds.), *Qu'est-ce que le care? Souci des autres, sensibilité, responsabilité* (Paris: Payot, 2009); Marie Garrau and Alice Le Goff, *Care, justice et dépendance. Introduction aux théories du care* (Paris: PUF, 2010); Pascale Molinier, *Le Care monde* (Lyon: ENS Éditions, 2018).

38 Annemarie Mol, *The Logic of Care: Health and the Problem of Patient Choice* (Abingdon and New York: Routledge, 2008).

39 Ibid., p. 35.

40 Joan C. Tronto, *Moral Boundaries: A Political Argument for an Ethic of Care* (New York: Routledge, 1993).

41 Berenice Fisher and Joan C. Tronto, 'Toward a feminist theory of caring', in Emily K. Abel and Margaret K. Nelson (eds.), *Circles of Care* (New York: SUNY Press, 1990), p. 40.

42 Antoine Hennion, 'Paying attention: What is tasting wine about?', in Ariane Barthoin Antal, Michael Hutter and David Stark (eds.), *Moments of Valuation: Exploring Sites of Dissonance* (Oxford: Oxford University Press, 2015), pp. 37–56.

Chapter 3: Attention

1 This scene comes from an investigation into the asset management of water networks coordinated by Daniel Florentin, in which one of us participated, including during this observation. See Daniel Florentin and Jérôme Denis, *Gestion patrimoniale des réseaux d'eau et d'assainissement en France*, Caisse des Dépôts-Institut pour la recherche et Banque des territoires, 2019.

2 We are here drawing, in abbreviated form, on a scene from the now classic work by Julian E. Orr on photocopier maintenance technicians: Julian E. Orr, *Talking about Machines: An Ethnography of a Modern Job* (Ithaca, NY: Cornell University Press, 1996), pp. 23–35.

3 We are reconstructing this scene from Tiziana Beltrame's research into museums: see Tiziana N. Beltrame, 'De la muséographie au bruit de fond biologique des collections', *Techniques & Culture*, no. 68, 2017, pp. 162–77.

4 Jonathan Waldman, *Rust: The Longest War* (New York: Simon & Schuster, 2015).

5 Steven J. Jackson, 'Speed, time, infrastructure: Temporalities of breakdown, maintenance, and repair', in Judy Wajcman and Nigel Dodd (eds.), *The Sociology of Speed: Digital, Organizational, and Social Temporalities* (Oxford: Oxford University Press, 2016), pp. 169–85.

6 Alexandra Bidet, *L'Engagement dans le travail. Qu'est-ce que le vrai boulot?* (Paris: PUF, 2011), pp. 280–90.

NOTES TO PP. 70–76

7 James J. Gibson, *The Ecological Approach to Visual Perception* (London: Routledge, 1986).

8 We will not here be going into the details of Gibson's thinking, which in any case has provoked countless commentaries. As well as his own work, we recommend Christian Bessy and Francis Chateauraynaud, *Experts et faussaires. Pour une sociologie de la perception* (Paris: Métailié, 1995), pp. 267–73; Philippe Descola, *Beyond Nature and Culture*, translated by Janet Lloyd (Chicago, IL: University of Chicago Press, 2014); pp. 186–90; and Donald Norman, 'Affordance, convention and design', *Interactions*, vol. 6, no. 3, 1999, pp. 38–43.

9 Orr, *Talking about Machines*.

10 Maria Puig de la Bellacasa, *Matters of Care: Speculative Ethics in More Than Human Worlds* (Minneapolis, MN: University of Minnesota Press, 2017).

11 This policy draws directly on the so-called 'broken windows theory'. See Joe Austin, *Taking the Train: How Graffiti Art Became an Urban Crisis in New York City* (New York: Columbia University Press, 2001); Ronald Kramer, 'Political elites, "broken windows", and the commodification of urban space', *Critical Criminology*, vol. 20, no. 3, 2012, pp. 229–48; Hunter Shobe and Tiffany Conklin, 'Geographies of graffiti abatement: Zero tolerance in Portland, San Francisco, and Seattle', *The Professional Geographer*, vol. 70, no. 4, 2018, pp. 624–32, on the situation in the United States; and Julie Vaslin, *Esthétique propre. La mise en administration des graffitis à Paris de 1977 à 2017*, PhD dissertation, Université Lyon 2, 2017, on Paris. The following descriptions are taken from an ethnographical project we carried out looking at municipal departments and graffiti removal companies between 2016 and 2019, the results of which we presented in part in Jérôme Denis and David Pontille, 'Maintenance epistemology and public order: Removing graffiti in Paris', *Social Studies of Science*, vol. 51, no. 2, 2021, pp. 233–58.

12 Antoine Hennion, 'The work to be made: The art of touching', in Bruno Latour and Christophe Leclercq (eds.), *Reset Modernity* (Cambridge, MA: The MIT Press, 2016), pp. 208–14.

13 Tim Dant, 'The "pragmatic" of material interaction', *Journal of Consumer Culture*, vol. 8, no. 1, 2008, pp. 11–33; and 'The work of repair: Gesture, emotion and sensual knowledge', *Sociological Research Online*, vol. 15, no, 3, 2010.

14 Johan M. Sanne, 'Making matters speak in railway maintenance', in Monika Büscher, Dawn Goodwin and Jessica Mesman (eds.), *Ethnographies of Diagnostic Work: Dimensions of Transformative Practice* (Basingstoke: Palgrave Macmillan, 2010), pp. 54–72.

15 Gilles Deleuze and Félix Guattari, *A Thousand Plateaus: Capitalism and*

Schizophrenia, translated by Brian Massumi (Minneapolis, MN and London: University of Minnesota Press, 1987), p. 409.

16 By paying due attention to these continuous material modulations, all those who practise maintenance can in their different ways endorse Karen Barad's analyses of what she calls 'intra-action', as well as Jane Bennett's reflections on the force of material vibrations in the composition of the world. See Karen Barad, *Meeting the Universe Halfway: Quantum Physics and the Entanglement of Matter and Meaning* (Durham, NC: Duke University Press, 2007); Jane Bennett, *Vibrant Matter: A Political Ecology of Things* (Durham, NC: Duke University Press, 2010).

17 Carlo Ginzburg, 'Clues: Roots of a scientific paradigm', *Theory and Society*, Vol. 7, no. 3, 1979, pp. 273–88.

18 Carlo Ginzburg and Anna Davin, 'Morelli, Freud and Sherlock Holmes: Clues and scientific method', *History Workshop*, no. 9, 1980, pp. 5–36 (p. 29).

19 Ginzburg, 'Clues', p. 281.

20 We have focused on the particular force of the maintenance record and its panoply of graphic signs in Jérôme Denis and David Pontille, 'Une écriture entre ordre et désordre. Le relevé de maintenance comme description normative', *Sociologie du travail*, vol. 56, no. 1, 2014, pp. 83–102.

21 Francis Chateauraynaud, 'Vigilance et transformation, présence corporelle et responsabilité dans la conduite des dispositifs techniques', *Réseaux*, no. 85, 1997, pp. 101–27.

22 Maurice Merleau-Ponty, *Phenomenology of Perception*, translated by Colin Smith (London and New York: Routledge, 2002).

23 Chateauraynaud, 'Vigilance et transformation', p. 120.

24 Sanne, 'Making matters speak in railway maintenance'.

25 Chateauraynaud, 'Vigilance et transformation', p. 121.

26 Vinciane Despret uses the expression '*faire connaissance*' ('getting to know') to underline the originality of certain ways of scientifically studying animals – recalcitrant beings by definition – that refuse to 'speak' in laboratory research situations. Getting to know each other involves the researcher listening, letting phenomena be expressed outside the purely deductive frameworks of modernist science, and being transformed by the relationship thus established. See Vinciane Despret, 'The enigma of the raven', *Angelaki: Journal of the Theoretical Humanities*, vol. 20, no. 2, Philosophical ethology II, 2015, pp. 57–72.

27 See Nikhil Anand, *Hydraulic City: Water and the Infrastructures of Citizenship in Mumbai* (Durham, NC: Duke University Press, 2017); Stephen Graham and Nigel Thrift, 'Out of order: Understanding repair and maintenance', *Theory, Culture & Society*, vol. 24, no. 3, 2007, pp. 1–25; Stephen Graham (ed.), *Disrupted Cities* (New York: Routledge, 2010), pp. 55–68; Stephen Graham and

NOTES TO PP. 88–94

Colin McFarlane (eds.), *Infrastructural Lives: Urban Infrastructure in Context* (Abingdon: Routledge, 2014); and Idalina Baptista, 'Electricity services always in the making: Informality and the work of infrastructure maintenance and repair in an African city', *Urban Studies*, vol. 56, no. 3, 2019, pp. 510–25.

28 This will be the subject of our last chapter, in which we will examine the conflicts encountered in certain areas of maintenance, and in particular the recent history of the movement for the defence of a 'right to repair'.

29 Yves Citton, *The Ecology of Attention*, translated by Barnaby Norman (Cambridge: Polity, 2017).

30 See the proposal in Estelle Zhong Mengual and Baptiste Morizot, *Esthétique de la rencontre. L'énigme de l'art contemporain* (Paris: Seuil, 2018).

31 See Antoine Hennion, 'Une sociologie des attachements. D'une sociologie de la culture à une pragmatique de l'amateur', *Sociétés*, no. 85, 2004, pp. 9–24; and 'Réflexivités. L'activité de l'amateur', *Réseaux*, vol. 153, no. 1, 2009, pp. 55–78.

32 These are the very people whom Baptiste Morizot mocks in his introduction to *Ways of Being Alive*, translated by Andrew Brown (Cambridge: Polity, 2022).

33 Cornelia Hummel and David Desaleux, *Mustang. La mécanique de la passion* (Lyon: Libel, 2018).

34 Ibid., pp. 65–95.

Chapter 4: Encounters

1 Tim Dant, *Materiality and Society* (New York: Open University Press, 2005), p. 115.

2 Tim Dant, 'The "pragmatic" of material interaction', *Journal of Consumer Culture*, vol. 8, no. 1, 2008, pp. 11–33.

3 Regarding scientific activity, Isabelle Stengers and Bruno Latour have also highlighted the importance of accepting the recalcitrance of the things studied, their capacity to oppose, or even rebel against, the expectations of researchers and to force them to revise their practices and their assumptions. See Bruno Latour, 'When things strike back: A possible contribution of "science studies" to the social sciences', *British Journal of Sociology*, vol. 51, no. 1, 2000, pp. 107–23; Isabelle Stengers, *The Invention of Modern Science*, translated by Daniel W. Smith (Minneapolis, MN and London: University of Minnesota Press, 2000). More generally, on the vitality of things, their specific capacity to manifest themselves according to their trajectories and their propensity to persist both with and without any human intervention, see Jane Bennett, *Vibrant Matter: A Political Ecology of Things* (Durham, NC: Duke University Press, 2010).

4 Matthew B. Crawford, *Shop Class as Soulcraft: An Inquiry into the Value of Work* (London: Penguin, 2009).

NOTES TO PP. 95–99

5 See previous chapter, and Julian E. Orr, *Talking about Machines: An Ethnography of a Modern Job* (Ithaca, NY: Cornell University Press, 1996).

6 Pierre Fournier, *Travailler dans le nucléaire. Enquête au cœur d'un site à risque* (Paris: Armand Colin, 2013); Marie-Aurore Ghis Malfilatre and Philippe Billard, 'Sous-traitance des risques: La maintenance de l'industrie nucléaire', in Annie Thébaud-Mony (ed.), *Les Risques du travail* (Paris: La Découverte, 2015), pp. 57–60.

7 Everett C. Hughes, 'Good people and dirty work', *Social Problems*, vol. 10, no. 1, 1962, pp. 3–11.

8 See Nicky Gregson, 'Performativity, corporeality and the politics of ship disposal', *Journal of Cultural Economy*, vol. 4, no. 2, 2011, pp. 137–56; Josh Lepawsky et al., 'Best of two worlds? Towards ethical electronics repair, reuse, repurposing and recycling', *Geoforum*, vol. 81, 2017, pp. 87–99.

9 Mohammad Rashidujjaman Rifat, Hasan Mahmud Prottoy and Syed Ishtiaque Ahmed, 'The breaking hand: Skills, care, and sufferings of the hands of an electronic waste worker in Bangladesh', *Proceedings of the 2019 CHI Conference on Human Factors in Computing Systems*, 2019, pp. 1–14.

10 Ibid., p. 23.

11 Fabrice Bourgeois et al., *Troubles musculo-squelettiques et travail. Quand la santé interroge l'organisation* (Paris: Anact, 2006).

12 Thierry Pillon, *Le Corps à l'ouvrage* (Paris: Stock, 2012).

13 Pascale Molinier, 'Et maintenant … que vais-je faire ? Incidences du progrès technique sur le travail des mécaniciens d'autobus', *Travailler*, no. 10, 2003, pp. 129–51.

14 The maintenance of underground networks raises an even more complex problem – the necessary opening of portions of the roadway, in a way adding one disassembly to another and making the collective organization of urban services an extremely delicate matter. For more on this aspect, see Jérôme Denis and Daniel Florentin, 'Des tuyaux qui comptent. Tournant patrimonial et renégociation des relations entre voirie et réseaux d'eau et d'assainissement', *Flux*, no. 126, 2022, pp. 32–46.

15 Lara Houston, 'The timeliness of repair', *Continent*, vol. 6, no. 1, 2017, pp. 51–7; and 'Mobile phone repair knowledge in downtown Kampala: Local and trans-local circulations', in Ignaz Strebel, Alain Bovet and Philippe Sormani (eds.), *Repair Work Ethnographies: Revisiting Breakdown, Relocating Materiality* (London: Palgrave Macmillan, 2019), pp. 129–60.

16 Nicolas Nova and Anaïs Bloch, *Dr. Smartphone: An Ethnography of Mobile Phone Repair Shops* (Geneva: Idpure, 2020).

17 We have described this scene in detail in Jérôme Denis and David Pontille,

NOTES TO PP. 100–104

'The dance of maintenance and the dynamics of urban assemblages: The daily (re)assemblage of Paris subway signs', in Strebel, Bovet and Sormani (eds.), *Repair Work Ethnographies*, pp. 161–85.

18 Blanca Callén and Tomás Sánchez Criado, 'Vulnerability tests: Matter of "care for matter" in e-waste practices', *Tecnoscienza*, vol. 6, no. 2, 2015, pp. 17–40.

19 Ibid.

20 Lepawsky et al., 'Best of two worlds?'

21 Among the many available studies, see Nicky Gregson et al., 'Following things of rubbish value: End-of-life ships, "chock-chocky" furniture and the Bangladeshi middle class consumer', *Geoforum*, vol. 41, no. 6, 2010, pp. 846–54; Uli Beisel and Tillmann Schneider, 'Provincialising waste: The transformation of Ambulance Car 7/83–2 to Tro-Tro Dr. Jesus', *Environment and Planning D*, vol. 30, no. 4, 2012, pp. 639–54; Jamie Cross and Declan Murray, 'The afterlives of solar power: Waste and repair off the grid in Kenya', *Energy Research & Social Science*, vol. 44, 2018, pp. 100–9.

22 Marianne de Laet and Annemarie Mol, 'The Zimbabwe bush pump: Mechanics of a fluid technology', *Social Studies of Science*, vol. 30, no. 2, 2000, pp. 225–63.

23 This is the type of technology studied by Madeleine Akrich. By following the interrupted trajectory of a photovoltaic lighting kit from France to Africa, she shows that one of the reasons for the failure of this 'technology transfer' was precisely that the kit was designed in such a way as to prevent any intervention by local technicians (Madeleine Akrich, 'The de-scription of technical objects', in Wiebe E. Bijker and John Law (eds.), *Shaping Technology/Building Society: Studies in Sociotechnical Change* [Cambridge, MA: The MIT Press, 1992], pp. 205–24). In chapter 7, we will come back to the conflicts which sometimes set the designers of technologies against those who assume responsibility for their maintenance.

24 Hilary Sample, *Maintenance Architecture* (Cambridge, MA: The MIT Press, 2018).

25 Steward Brand, *How Buildings Learn: What Happens after They're Built* (New York: Viking Penguin, 1994).

26 This is clear from Ignaz Strebel's investigations, carried out with maintenance staff in collective housing buildings. See Ignaz Strebel, 'The living building: Towards a geography of maintenance work', *Social & Cultural Geography*, vol. 12, no. 3, 2011, pp. 243–62.

27 Robert Shaw, 'Cleaning up the streets: Newcastle-upon-Tyne's night-time neighbourhood services team', in Stephen Graham and Colin McFarlane (eds.), *Infrastructural Lives: Urban Infrastructure in Context* (Abingdon: Routledge, 2014), pp. 174–96.

NOTES TO PP. 107–115

28 We must make a caveat here, and recognize that all removers (like all maintainers) do not act in the same way, and not all are equally careful. There are people in all domains who are said to carry out 'bad' maintenance, in particular because they are careless and do not take 'good' care of things. We will address some of these criticisms later in this book.

29 Tim Dant, 'The work of repair: Gesture, emotion and sensual knowledge', *Sociological Research Online*, vol. 15, no. 3, 2010.

30 Steven J. Jackson and Lara Houston, 'The poetics and political economy of repair', in Jeremy Swartz and Janet Wasko (eds.), *Media: A Transdisciplinary Inquiry* (Chicago, IL: Intellect Books, 2021), pp. 244–64.

31 Alain Rey, *Dictionnaire historique de la langue française* (Paris: Le Robert, 1998).

32 On this point, see Antoine Hennion and Pierre Vidal-Naquet on home help: Antoine Hennion and Pierre Vidal-Naquet, 'La contrainte est-elle compatible avec le *care*? Le cas de l'aide et du soin à domicile', *Alter*, vol. 9, no. 3, 2015, pp. 207–21.

33 Andrew Pickering, *The Mangle of Practice: Time, Agency, and Science* (Chicago, IL: University of Chicago Press, 1995); Andrew Pickering, 'The robustness of science and the dance of agency', in Léna Soler et al. (eds.), *Characterizing the Robustness of Science: Boston Studies in the Philosophy of Science* (Dordrecht: Springer Netherlands, 2012), pp. 317–27.

34 In certain scientific fields, the different parts played are explicitly acknowledged and even displayed in articles where the names of the instruments and the names of the researchers coexist. See David Pontille, *Signer ensemble. Contribution et évaluation en sciences* (Paris: Économica, 2016).

35 Tim Ingold, *Making: Anthropology, Archaeology, Art and Architecture* (Abingdon: Routledge, 2013).

36 Gilles Deleuze, *Difference and Repetition*, translated by Paul Patton, 2nd edn (London and New York: Bloomsbury, 2014).

Chapter 5: Time

1 This is, of course, a pseudonym.

2 Lazar Kunstmann, *Urban eXperiment* (Paris: Uqbar, 2018).

3 This 'branch' of UX organized, for example, film festivals, including the one held at the intersection of several old quarries and the foundations of the Palais de Chaillot, the existence of which was made public in 2004 following an anonymous call to the police.

4 Among the articles published at the time, see F. Gouaillard, 'Le Panthéon habité en secret', *Le Parisien*, 14 October 2016; Clarisse Fabre, 'Aux intrus, la patrie très énervée', *Le Monde*, 24 November 2007; E. Boyer King, 'Undercover

NOTES TO PP. 116–127

restorers fix Paris landmark's clock', *The Guardian*, 26 November 2007; Anne Vidalie and Anne-Laure Pham, 'Sous les pavés, les explorateurs', *L'Express*, 27 March 2008; A. Sage, 'Underground terrorists with a mission to save city's neglected heritage', *The Times*, 29 September 2007.

5 Kunstmann, *Urban eXperiment*, p. 55.

6 To get an idea of this, see – in addition to the photos in his book – Kunstmann's film *Panthéon, mode d'emploi*: https://vimeo.com/51365068.

7 Jon Lackman, 'The new French hacker-artist underground', *Wired*, 28 January 2012.

8 He did this in the interview given to Jon Lackman for the magazine *Wired* (ibid.), in the work he himself published (Kunstmann, *Urban eXperiment*), and at a conference given at the prestigious Long Now Foundation, at the invitation of Alexander Rose and Stewart Brand, two of its founders, available online at: https://longnow.org/seminars/02012/nov/13/preservation-without-permission -paris-urban-experiment.

9 We will be returning to the debates on heritage conservation in the next chapter.

10 Kunstmann, *Urban eXperiment*, p. 31.

11 These are the terms of the title of one of Kunstmann's chapters (*Urban eXperiment*).

12 We thank Roman Solé-Pomies who suggested this expression, which underlines the proper value of this alternative clock, set in place when Pascal Monnet restored it in his own way.

13 Antoine Hennion, 'La restauration, un atelier de l'histoire', in Noémie Etienne and Léonie Hénaut (eds.), *L'Histoire à l'atelier. La restauration des œuvres d'art, XVIIIe–XXIe siècles* (Lyon: Presses Universitaires de Lyon, 2012), pp. 399–412.

14 Ibid., p. 399.

15 Kunstmann, *Urban eXperiment*, p. 203.

16 François Hartog, *Regimes of Historicity: Presentism and Experiences of Time*, translated by Saskia Brown (New York: Columbia University Press, 2015).

17 To find out more on this subject, see Jonnet Middleton's discussion – both political and analytical – in 'Mending', in Kate Fletcher and Mathilda Tham (eds.), *Routledge Handbook of Sustainability and Fashion* (Abingdon: Routledge, 2014), pp. 262–74.

18 Nicky Gregson, Alan Metcalfe and Louise Crewe, 'Practices of object maintenance and repair: How consumers attend to consumer objects within the home', *Journal of Consumer Culture*, vol. 9, no. 2, 2009, pp. 248–72.

19 Nicky Gregson and her colleagues show, however, that not everyone shares this opinion in the households concerned and that the 'relaxed' feel of a prolongation can be severely criticized and give rise to quarrels. We will return to the question of such disputes in the next two chapters.

NOTES TO PP. 127–140

20 As Nicky Gregson and her colleagues point out, not allowing objects in the domestic sphere to wear out too quickly is also frequently a matter of reputation among friends and local residents.

21 David Edgerton, 'Creole technologies and global histories: Rethinking how things travel in space and time', *Journal of History of Science and Technology*, vol. 1, no. 1, 2007, pp. 3–31.

22 Lara Houston, 'The timeliness of repair', *Continent*, vol. 6, no. 1, 2017, pp. 51–7.

23 David Edgerton, *The Shock of the Old: Technology and Global History since 1900* (Oxford: Oxford University Press, 2007).

24 Uli Beisel and Tillmann Schneider, 'Provincialising waste: The transformation of Ambulance Car 7/83–2 to Tro-Tro Dr. Jesus', *Environment and Planning D*, vol. 30, no. 4, 2012, pp. 639–54; Edgerton, 'Creole technologies and global histories'.

25 This is the subject of chapter 7.

26 Alexei Yurchak, 'Bodies of Lenin: The hidden science of Communist sovereignty', *Representations*, vol. 129, no. 1, 2015, pp. 116–57.

27 This 'turn towards the user' has been common in other sectors, too, and is considered one of the main axes of what, at the end of the 1980s, was called the 'modernization' of public services. Isaac Joseph and Gilles Jeannot, *Métiers du public. Les compétences de l'agent et l'espace de l'usager* (Paris: CNRS, 1995); Jean-Marc Weller, 'La modernisation des services publics par l'usager. Une revue de la littérature (1986–1996)', *Sociologie du travail*, vol. 40, no. 3, 1998, pp. 365–92.

28 Alexei Yurchak, 'Communist proteins: Lenin's skin, astrobiology, and the origin of life', *Kritika*, vol. 20, no. 4, 2019, pp. 683–715.

29 We encountered these situations several times during our research. For several months, the Campo Formio station, for example, was equipped with a sign displaying the name of the station in Comic Sans font. One day, we also came across a made-up panel on the wall of the station, representing somewhat haphazardly official forms of signage, including arrows and circles for the number of lines, with the typography reproduced manually. This was doubtless a facsimile composed in an emergency in response to a disappearance. For more details, see Jérôme Denis and David Pontille, *Petite sociologie de la signalétique. Les coulisses des panneaux du métro* (Paris: Presses des Mines, 2010).

30 Yurchak, 'Bodies of Lenin', p. 116.

31 There was thus never any question of producing a facsimile resembling effigies, those 'second bodies of the king' whose importance in monarchical regimes undergoing secularization was demonstrated by Kantorowicz. See Ernst Kanotorowicz, *The King's Two Bodies: An Essay in Mediaeval Political Theology* (Princeton, NJ: Princeton University Press, 1957).

NOTES TO PP. 141–150

32 Laurent Olivier, *The Dark Abyss of Time: Archaeology and Memory*, translated by Arthur Greenspan (Lanham, MD: AltaMira Press, 2011), pp. 87–94.

33 Olivier also draws on the work of historian Sarah Farmer, *Martyred Village: Commemorating the 1944 Massacre at Oradour-sur-Glane* (Berkeley, CA: University of California Press, 2000).

34 Olivier, *The Dark Abyss of Time*, p. 57.

35 Michel de Certeau, *The Writing of History*, translated by Tom Conley (New York: Columbia University Press, 1992).

36 Bruno Latour, *We Have Never Been Modern*, translated by Catherine Porter (Cambridge, MA: Harvard University Press, 1993); Isabelle Stengers, *The Invention of Modern Science*, translated by Daniel W. Smith (Minneapolis, MN and London: University of Minnesota Press, 2000).

37 This does not undermine – quite the contrary – the importance of the diplomatic aspect of the dance of maintenance that we highlighted in the previous chapter. The case of Lenin's body, in particular, is a perfect illustration of the negotiations that it is necessary to invent directly with the materials in order to succeed in producing a form of eternity.

38 The report by Alexander Rose, co-founder of Long Now Foundation, describing the latest ceremony, is somewhat anecdotal; available online at: https://longnow.org/ideas/02019/09/11/long-term-building-in-japan. More precise and better documented details can be found in the scholarly literature, notably Felicia G. Bock, 'The rites of renewal at Ise', *Monumenta Nipponica*, vol. 29, no. 1, 1974, pp. 55–68; Cassandra Adams, 'Japan's Ise shrine and its thirteen-hundred-year-old reconstruction tradition', *Journal of Architectural Education*, vol. 52, no. 1, 1998, pp. 49–60; John K. Nelson, *Enduring Identities: The Guise of Shinto in Contemporary Japan* (Honolulu, HI: University of Hawaii Press, 2000); Dominic Mciver Lopes, 'Shikinen Sengu and the ontology of architecture in Japan', *The Journal of Aesthetics and Art Criticism*, vol. 65, no. 1, 2007, pp. 77–84.

39 The Ise complex is mainly devoted to the worship of Amaterasu, the sun goddess.

40 Moritz F. Fürst, '"A good enough fix": Repair and maintenance in librarians' digitization practice', in Ignaz Strebel, Alain Bovet and Philippe Sormani (eds.), *Repair Work Ethnographies: Revisiting Breakdown, Relocating Materiality* (London: Palgrave Macmillan, 2019), pp. 61–87.

41 Ibid., p. 79.

42 Fernando Domínguez Rubio, *Still Life: Ecologies of the Modern Imagination at the Art Museum* (Chicago, IL: University of Chicago Press, 2020).

43 Fernando Domínguez Rubio, 'On the discrepancy between objects and things', *Journal of Material Culture*, vol. 21, no. 1, 2016, pp. 59–86.

NOTES TO PP. 155–166

44 We will return to this question in more detail in the next chapter.

45 Kunstmann, *Urban eXperiment*, p. 39.

46 See the Euronews report, available online at: https://www.youtube.com/watch?v=-bR5yshYtd5E.

47 Elizabeth Pye, *Caring for the Past: Issues in Conservation for Archeology and Museums* (London: James & James Ltd, 2001); David Lowenthal, *The Past Is a Foreign Country (Revisited)* (Cambridge: Cambridge University Press, 2015); Jean-Pierre Cometti, *Conserver/Restaurer. L'œuvre d'art à l'époque de sa préservation technique* (Paris: Gallimard, 2016).

48 Caitlin DeSilvey, *Curated Decay: Heritage beyond Saving* (Minneapolis, MN: University of Minnesota Press, 2017).

49 Ibid., p. 16.

50 Karen Barad, 'Posthumanist performativity: Toward an understanding of how matter comes to matter', *Signs*, vol. 28, no. 3, 2003, pp. 801–31; Jane Bennett, 'The force of things: Steps toward an ecology of matter', *Political Theory*, vol. 32, no. 3, 2004, pp. 347–72; Tim Ingold, 'Materials against materiality', *Archaeological Dialogues*, vol. 14, no. 1, 2007, pp. 1–16.

51 There have been heated debates in environmentalist circles over this idea of the wilderness, debates that have drawn especially on the critique in John Baird Callicott and Michael P. Nelson (eds.), *The Great New Wilderness Debate* (Athens, GA: University of Georgia Press, 1998). We will take a closer look at this question in the next chapter.

52 Georg Simmel, 'The ruin', in Georg Simmel et al., *Essays on Sociology, Philosophy and Aesthetics*, edited by Kurt H. Wolff (New York: Harper and Row, 1965), pp. 259–66.

53 Marisa L. Cohn, 'Convivial decay: Entangled lifetimes in a geriatric infrastructure', *Proceedings of ACM Conference on Computer Supported Collaborative Work*, 2016; and '"Lifetime issues": Temporal relations of design and maintenance', *Continent*, vol. 6, no. 1, 2017, pp. 4–12.

54 Cohn, 'Convivial decay'.

55 Ibid., p. 9.

56 Émilie Gomart and Antoine Hennion, 'A sociology of attachment: Music amateurs, drug users', *The Sociological Review*, vol. 47, no. 1, 1999, pp. 220–47; Antoine Hennion, 'Attachments, you say? … How a concept collectively emerges in one research group', *Journal of Cultural Economy*, vol. 10, no.1, 2017, pp. 112–21; and 'Enquêter sur nos attachements. Comment hériter de William James?', *SociologieS*, 2015, avtps://doi.org/10.4000/sociologies.4953.

57 Joan C. Tronto, *Moral Boundaries: A Political Argument for an Ethic of Care* (New York: Routledge, 1993). For a recent discussion of the ambiguities in

NOTES TO PP. 166–171

the notion of care, see Vincent Duclos and Tomás Sanchéz Criado, 'Care in trouble: Ecologies of support from below and beyond', *Medical Anthropology Quarterly*, vol. 34, no. 2, 2020, pp. 153–73.

58 Başak Saraç-Lesavre, 'Desire for the "worst": Extending nuclear attachments in southeastern New Mexico', *Environment and Planning D*, vol. 38, no. 4, 2020, pp. 753–71; and 'Faire avec les restes de l'âge nucléaire', *Revue d'Anthropologie des Connaissances*, vol. 14, no. 4, 2020.

59 Bernadette Bensaude-Vincent, *Temps-paysage. Pour une écologie des crises* (Paris: Le Pommier, 2021).

60 To give just two examples, we can think of the question of limited resources in the energy field and the specific temporal constraints caused by the need to make restricted use of them, and also reflect on so-called 'digital' technologies, in particular the archiving of constantly updated data, a process that represents unprecedented conservation challenges; here, the question of time doubtless assumes significantly different shapes from the four configurations that we have detailed here.

61 Bensaude-Vincent, *Temps-paysage*.

Chapter 6: Tact

1 Christian Bessy and Francis Chateauraynaud, *Experts et faussaires. Pour une sociologie de la perception* (Paris: Métailié, 1995). For a presentation of their main argument in English, see Christian Bessy and Francis Chateauraynaud, 'The dynamics of authentication and counterfeits in markets', *Historical Social Research / Historische Sozialforschung*, vol. 44, no. 1, 2019, pp. 136–59.

2 Bessy and Chateauraynaud, *Experts et faussaires*, p. 9.

3 Luc Boltanski and Ève Chiapello saw it as a paradigmatic example of the capacities of capitalism to overcome critique through an operation of constant recuperation: see Luc Boltanski and Ève Chiapello, *The New Spirit of Capitalism*, translated by Gregory Elliott (London: Verso, 2018).

4 See David Lowenthal, 'Authenticity? The dogma of self-delusion', in Mark Jones (ed.), *Why Fakes Matter: Essays on Problems of Authenticity* (London: British Museum Press, 1992), pp. 184–92; Sharon Zukin, 'Consuming authenticity', *Cultural Studies*, vol. 22, no. 5, 2008, pp. 724–48; Andrew Potter, *The Authenticity Hoax: How We Get Lost Finding Ourselves* (Melbourne and London: Scribe Publications, 2010); Jillian R. Cavanaugh and Shalini Shankar, 'Producing authenticity in global capitalism: Language, materiality, and value', *American Anthropologist*, vol. 116, no. 1, 2014, pp. 51–64.

5 Cornelia Hummel and David Desaleux, *Mustang. La mécanique de la passion* (Lyon: Libel, 2018).

NOTES TO PP. 173–190

6 Ibid., p. 69.

7 Some go so far as to adopt that most reprehensible of all practices, parking across two spaces to keep a sufficient distance between neighbouring cars and ensure the paint is not scratched by a clumsily opened door.

8 Hummel and Desaleux, *Mustang*.

9 Ibid.

10 William James, *Essays in Radical Empiricism*, edited by Frederick Burkhardt and Fredson Bowers (Cambridge, MA: Harvard University Press, 1976).

11 See Jean-Pierre Cometti, *Conserver/Restaurer. L'œuvre d'art à l'époque de sa préservation technique* (Paris: Gallimard, 2016); Fernando Domínguez Rubio, *Still Life: Ecologies of the Modern Imagination at the Art Museum* (Chicago, IL: University of Chicago Press, 2020); and Yaël Kreplak, 'Quelle sorte d'entité matérielle est une œuvre d'art? Le cas du Magasin de Ben', *Images Re-vues*, available online at: http://journals.openedition.org/imagesrevues/6396.

12 Jérôme Denis, Cornelia Hummel and David Pontille, 'Getting attached to a Classic Mustang: Use, maintenance and the burden of authenticity', *Journal of Material Culture*, 2022, vol. 27, no. 3, p. 259–79.

13 Albena Yaneva, 'How buildings "surprise": The renovation of the Alte Aula in Vienna', *Science Studies*, vol. 21, no. 1, 2001, pp. 8–28.

14 Étienne Souriau, *The Different Modes of Existence: Followed by On the Work to Be Made*, translated by Erik Beranek and Tim Howles (Minneapolis, MN: Univocal Publishing, 2015).

15 Thomas Yarrow, 'How conservation matters: Ethnographic explorations of historic building renovation', *Journal of Material Culture*, vol. 24, no. 1, 2019, pp. 3–21 (p. 12).

16 Anne Savalli, *Chartres, la lumière retrouvée. Chronique d'une restauration*, Kanari Films, 2016.

17 Aloïs Riegl, 'The modern cult of monuments: Its character and its origin', translated by Kurt W. Forster and Diane Ghirardo, *Oppositions*, no. 25, Fall 1982, pp. 21–51.

18 François Hartog, *Regimes of Historicity: Presentism and Experiences of Time*, translated by Saskia Brown (New York: Columbia University Press, 2015).

19 Jean-Michel Leniaud, *Les Archipels du passé. Le patrimoine et son histoire* (Paris: Fayard, 2002).

20 As we mentioned, there is a very extensive literature on this subject. For our incursions into this area, we are drawing mainly on: John Delafons, *Politics and Preservation: A Policy History of the Built Heritage 1882–1996* (London: Routledge, 1997); Sophia Labadi, 'World heritage, authenticity and post-authenticity: International and national perspectives', in Sophia Labadi and

Colin Long (eds.), *Heritage and Globalisation* (Abingdon: Routledge, 2010), pp. 66–84; Cornelius Holtorf, 'The heritage of heritage', *Heritage & Society*, vol. 5, no. 2, 2012, pp. 153–74; Emma Waterton and Steve Watson (eds.), *The Palgrave Handbook of Contemporary Heritage Research* (New York: Springer, 2015); David Lowenthal, *The Past Is a Foreign Country (Revisited)* (Cambridge: Cambridge University Press, 2015); Helaine Silverman, 'Heritage and authenticity', in Waterton and Watson (eds.), *The Palgrave Handbook of Contemporary Heritage Research*, pp. 69–88; Rodney Harrison, 'On heritage ontologies: Rethinking the material worlds of heritage', *Anthropological Quarterly*, vol. 91, no. 4, 2018, pp. 1365–83.

21 Lowenthal, *The Past Is a Foreign Country*.

22 Antoine Hennion, 'La restauration, un atelier de l'histoire', in Noémie Étienne and Léonie Hénaut (eds.), *L'Histoire à l'atelier. La restauration des œuvres d'art, XVIIIe–XXIe siècles* (Lyon: Presses Universitaires de Lyon, 2012), pp. 399–412.

23 Lowenthal, *The Past Is a Foreign Country*, p. 468.

24 Many aspects of Ruskin's and Morris's position seem to directly echo the 'post-preservationist' speculations of Caitlin DeSilvey that we discussed in the previous chapter. Surprisingly, however, she never mentions these authors.

25 Henry James, *A Little Tour in France*, available online at: https://www.gutenberg.org/cache/epub/2159/pg2159-images.html; see also Jean-Michel Leniaud, *Les Archipels du passé. Le patrimoine et son histoire* (Paris: Fayard, 2002).

26 Camillo Boito, 'Conservare o restaurare', in *Questioni pratiche di belli arti – Restauri, concorsi, legislazione, professione, insegnamento* (Milan: Librario della Real Casa, 1893).

27 Quoted by Morgane Poirier in 'La notion de réversibilité en conservation-restauration', in research notes published online at 'Tables de travail', available online at: https://tablesdetravail.hypotheses.org/721.

28 Françoise Choay, *L'Allégorie du patrimoine* (Paris: Seuil, 1996).

29 Hartog, *Regimes of Historicity*.

30 Eugène-Emmanuel Viollet-le-Duc, 'Restauration', in *Dictionnaire raisonné de l'architecture française du XIe au XVIe siècle, 1854–1868*, vol. 8 (Paris: Bance-Morel, 1866), p. 14.

31 Leniaud, *Les Archipels du passé*.

32 Ibid.

33 For a meticulous genealogy of the notion, see Ludovic Roudet, *L'Intervention minimale en conservation-restauration des biens culturels. Exploration d'une notion*, master's dissertation, Université Paris I, Panthéon-Sorbonne, 2007.

34 Icomos, *International Charter for the Conservation and Restoration of Monuments and Sites (The Venice Charter 1964)*, Second International Congress of Architects

NOTES TO PP. 196–201

and Technicians of Historic Monuments, Venice, 1964, available online at: https://www.icomos.org/images/DOCUMENTS/Charters/venice_e.pdf.

35 Roudet (*L'Intervention minimale*) traces its first appearance back to article 18 of the first Code of Ethics and Guidance for Practice, first issued in 1986, jointly published by CAPC and the Canadian Association for Conservation of Cultural Property (CAC), available online at: https://capc-acrp.ca/en/what-is-conservation/publications/code-of-ethics-and-guidance-for-practice.

36 Iccrom and Icomos, *Management Guidelines for World Cultural Heritage Sites*, 1993, available online at: https://www.iccrom.org/sites/default/files/2018-02/1998_feilden_management_guidelines_eng_70071_light_0.pdf (p. 77; our emphasis).

37 https://www.icomos.org/charters/nara-e.pdf.

38 This is the main argument put forward by Caroline Villers in a highly influential article. See Caroline Villers, 'Post minimal intervention', *The Conservator*, vol. 28, no. 1, 2004, pp. 3–10.

39 Donna J. Haraway, 'Modest_ witness@second_millennium', the first chapter in her *Second Millennium.FemaleMan© Meets OncoMouseTM. Feminism and Technoscience* (New York and London: Routledge, 1997), p. 26.

40 Rotor, *Usus/usures. État des lieux – How things stand*, available online at: https://rotordb.org/sites/default/files/2019-06/usus_usures_Rotor.pdf (p. 55).

41 See in particular the artwork *LANDING*, developed in the context of the transformation of a huge open-air landfill into a park at Fresh Kills, New York: https://freshkillspark.org/os-art/landing.

42 Catherine Larrère, 'Environmental ethics: Respect and responsibility', in Aurélie Choné, Isabelle Hajek, Philippe Hamman (eds.), *Rethinking Nature: Challenging Disciplinary Boundaries* (Abingdon and New York: Routledge, 2017), pp. 54–83.

43 William Cronon explains how essayist Fredrick Turner himself participated in constructing this myth by imbuing the spirituality of Muir and Thoreau with a national dimension and by making the wilderness the 'quintessence of Americanness'. See William Cronon, 'The trouble with wilderness, or getting back to the wrong nature', in William Cronon (ed.), *Uncommon Ground: Rethinking the Human Place in Nature* (New York: W. W. Norton, 1995), p. 176.

44 Aldo Leopold, *A Sand County Almanac, and Sketches Here and There* (Oxford: Oxford University Press, 1949).

45 In particular Susan Flader and, in France, Catherine and Raphaël Larrère.

46 Aldo Leopold, 'The popular wilderness fallacy', in Susan L. Flader and John Baird Callicott (eds.), *The River of the Mother of God* (Madison, WI: University of Wisconsin Press, 1991), pp. 49–52; John Baird Callicott, *In Defense of the*

NOTES TO PP. 201–205

Land Ethic: Essays in Environmental Philosophy (Albany, NY: State University of New York Press, 1989).

47 Leopold, 'The popular wilderness fallacy'.

48 Cronon, 'The trouble with wilderness', pp. 69–90.

49 See for example Marina Mies and Vandana Shiva, *Ecofeminism* (London: Zed Books, 1999).

50 Philippe Descola, *Beyond Nature and Culture*, translated by Janet Lloyd (Chicago, IL: University of Chicago Press, 2014); Eduardo Viveiros de Castro, *Cannibal Metaphysics*, edited and translated by Peter Skafish (Minneapolis, MN: University of Minnesota Press, Univocal Publishing, 2014).

51 Lynn White Jr lays the blame for this separation on Christianity, which, by seeing human beings as a divine creation, authorizes them to exploit nature as a resource. Lynn White, 'The historical roots of our ecologic crisis', *Science*, no. 155, 1967, pp. 1203–7.

52 In John Passmore's view, this fundamental separation stems more from the figure of human beings taming the forces of nature by the means of science – a figure present in particular in Bacon and Descartes. See John Passmore, *Man's Responsibility for Nature: Ecological Problems and Western Tradition* (New York: Scribner, 1974).

53 Callicott, *In Defense of the Land Ethic*, p. 216.

54 For a recent overview of the question of the permanent transformations of living things, see Emanuele Coccia, *Metamorphoses*, translated by Robin Mackay (Cambridge: Polity, 2012).

55 Carolyn Merchant, *Reinventing Eden: The Fate of Nature in Western Culture* (New York: Routledge, 2003).

56 Callicott, *In Defense of the Land Ethic*.

57 Cronon, 'The trouble with wilderness'.

58 Aldo Leopold, 'The farmer as a conservationist', *A Sand County Almanac, and Sketches Here and There* (Oxford: Oxford University Press, 1949), pp. 425–37.

59 Carolyn Merchant uses this term to imagine positions that ensure the conditions for cohabitation, without presupposing that the action of humans is necessarily harmful to other entities present, but without denying the fact that the environment can be dangerous for humans. See Merchant, *Reinventing Eden*.

60 Kirkpatrick Sale, *Dwellers in the Land: The Bioregional Vision* (San Francisco, CA: Sierra Club Books, 1985); Mathias Rollot and Marin Schaffner, *Qu'est-ce qu'une biorégion?* (Marseilles: Wildproject, 2021).

61 Benedikte Zitouni et al., *Terres des villes. Enquêtes potagères aux premières saisons du 21e siècle* (Paris: L'Éclat, 2018).

NOTES TO PP. 205–210

62 Perrine Hervé-Gruyer and Charles Hervé-Gruyer, *Permaculture. Guérir la terre, nourrir les hommes* (Arles: Actes Sud, 2015).

63 Gilles Clément retraces his career and describes in detail the work he carried out at the André Citroën park with the Paris gardeners team; in Gilles Clément, 'Jardins en mouvement, friches urbaines et mécanismes de la vie', *Journal d'agriculture traditionnelle et de botanique appliquée*, vol. 39, no. 2, 1997, pp. 157–75.

64 Ibid., p. 158.

65 Ibid., pp. 159–60.

66 Gilles Clément, 'La friche apprivoisée', *Urbanisme*, no. 209, 1985, pp. 91–5.

67 Raphaël Mathevet and Arnaud Béchet, *Politiques du flamant rose. Vers une écologie du sauvage* (Marseilles: Wildproject, 2020).

68 Ibid., p. 110.

69 Ibid., p. 93.

70 Larrère, 'Environmental ethics'.

71 Raphaël Mathevet, *La Solidarité écologique. Ce lien qui nous oblige* (Arles: Actes Sud, 2012).

72 Vinciane Despret, *Our Grateful Dead: Stories of Those Left Behind* (Minneapolis, MN: University of Minnesota Press, 2021).

73 Maria Puig de la Bellacasa, *Matters of Care: Speculative Ethics in More Than Human Worlds* (Minneapolis, MN: University of Minnesota Press, 2017), p. 115.

74 Andrew Pickering, *The Mangle of Practice: Time, Agency, and Science* (Chicago, IL: University of Chicago Press, 1995); Tim Ingold, *Making: Anthropology, Archaeology, Art and Architecture* (London: Routledge, 2013). See our discussion in chapter 4.

75 Pickering, *The Mangle of Practice*.

76 Siân Jones and Thomas Yarrow, 'Crafting authenticity: An ethnography of conservation practice', *Journal of Material Culture*, vol. 18, no. 1, 2013, pp. 3–26.

77 Jérôme Denis and David Pontille, 'Maintenance epistemology and public order: Removing graffiti in Paris', *Social Studies of Science*, vol. 51, no. 2, 2021, pp. 233–58.

78 The many great studies in this area include: Judith Plant (ed.), *Healing the Wounds: The Promise of Ecofeminism* (Philadelphia: New Society Publishers, 1989); Val Plumwood, 'Nature, self, and gender: Feminism, environmental philosophy, and the critique of rationalism', *Hypatia*, vol. 6, no. 1, 1991, pp. 3–27; Carolyn Merchant, *Earthcare, Women and the Environment* (New York: Routledge, 1996); Catherine Larrère, 'L'écoféminisme. Féminisme écologique ou écologie féministe', *Tracés*, no. 22, 2012, pp. 105–21; Marina Mies and Vandana Shiva, *Ecofeminism* (London: Zed Books, 1999); Catherine

NOTES TO PP. 212–218

Larrère, 'La nature a-t-elle un genre? Variétés d'écoféminisme', *Cahiers du genre*, no. 2, 2015, pp. 103–25.

Chapter 7: Conflicts

1 https://fr.ifixit.com/News/41440/introducing-the-worlds-largest-medical-repair-database-free-for-everyone.
2 https://twitter.com/kwiens/status/1271134890872856577.
3 https://uspirg.org/news/usp/43000-call-ventilator-manufacturers-release-repair-information.
4 https://uspirg.org/news/usp/hospital-repair-professionals-just-let-us-fix-life-saving-devices-including-ventilators?_ga=2.44051492.1224110732.1590153584-1356731714.1590153584. [The exact wording of the letter is no longer available at this URL – Trans.]
5 http://frankshospitalworkshop.com.
6 https://www.fda.gov/media/113431/download.
7 Stephen L. McIntyre, 'The failure of Fordism: Reform of the automobile repair industry, 1913–1940', *Technology and Culture*, vol. 41, no. 2, 2000, pp. 269–99.
8 Alexandre Mallard, 'Être garagiste au temps de l'informatique', in Christian Bromberger and Denis Chevallier (eds.), *Carrières d'objets. Innovations et relances* (Paris: Éditions de la MSH, 1999), pp. 59–81; Emmanuelle Dutertre and Bernard Jullien, 'Les artisans de la réparation automobile face aux constructeurs. Vers l'affirmation d'un contre-modèle', *Revue d'anthropologie des connaissances*, vol. 9, no. 3, 2015, pp. 331–50.
9 Leah Chan Grinvald and Ofer Tur-Sinai, 'The right to repair: Perspectives from the United States', *Australian Intellectual Property Journal*, vol. 31, no. 2, 2020, pp. 98–110; Masayuki Hatta, 'The right to repair, the right to tinker, and the right to innovate', *Annals of Business Administrative Science*, vol. 19, no. 4, 2020, pp. 143–57.
10 One of the first is by Jason Koebler, who has drawn on the dossier and has since published several others. 'Why American farmers are hacking their tractors with Ukrainian firmware', https://www.vice.com. We shall be returning to this.
11 https://ec.europa.eu/environment/circular-economy, March 2020.
12 https://www.whitehouse.gov/briefing-room/statements-releases/2021/07/09/fact-sheet-executive-order-on-promoting-competition-in-the-american-economy.
13 https://www.apple.com/newsroom/2021/11/apple-announces-self-service-repair.
14 Jeanne Guien, *Obsolescences. Philosophie des techniques et histoire économique à l'épreuve de la réduction de la durée de vie des objets*, PhD dissertation, Université Panthéon-Sorbonne-Paris I, Paris, 2019.

NOTES TO PP. 219–226

15 Any readers curious to find out more about this still little-known story can read the dissertation by Jeanne Guien (see previous note), as well as the essential study by Giles Slade, *Made to Break: Technology and Obsolescence in America* (Cambridge, MA: Harvard University Press, 2006); see also Susan Strasser, *Waste and Want: A Social History of Trash* (New York: Metropolitan Books, 1999), in which the author discusses the question of obsolescence at length.

16 Markus Krajewski, 'The great lightbulb conspiracy', *IEEE Spectrum*, vol. 51, no. 10, 2014, pp. 56–61.

17 For a recent survey of market *agencements*, see Michel Callon, *Markets in the Making: Rethinking Competition, Goods, and Innovation*, translated by Olivia Custer, edited by Martha Poon (Brooklyn, NY: Zone Books, 2021).

18 Krajewski, 'The great lightbulb conspiracy', p. 60.

19 Jeanne Guien, *Le Consumérisme à travers ses objets* (Paris: Éditions Divergences, 2021).

20 Strasser, *Waste and Want*.

21 Slade, *Made to Break*, p. 25.

22 Joan C. Tronto, *Moral Boundaries: A Political Argument for an Ethic of Care* (New York: Routledge, 1993).

23 Franck Cochoy, *Une histoire du marketing. Discipliner l'économie de marché* (Paris: La Découverte, 1999); Sandrine Barrey, Franck Cochoy and Sophie Dubuisson-Quellier, 'Designer, packager, merchandiser. Trois professionnels pour une même scène marchande', *Sociologie du travail*, vol. 42, no. 3, 2000, pp. 457–82; Thibault Le Texier, *La Main visible des marchés. Une histoire critique du marketing* (Paris: La Découverte, 2022).

24 Henry Ford, *My Life and Work* (New York: Macmillan, 1922), available online at: http://library.manipaldubai.com/DL/My_Life_and_Work.pdf (p. 59).

25 Ibid.

26 Paul M. Mazur, *American Prosperity: Its Causes and Consequences* (New York: The Viking Press, 1928), pp. 92–3.

27 Hatta, 'The right to repair'.

28 Vance Packard, *The Hidden Persuaders* (London: Lowe & Brydone Ltd., 1957); and *The Waste Maker*s (New York: Ig Publishing, 1960).

29 Jean Baudrillard, *The System of Objects*, translated by James Benedict (London: Verso, 1996); Marshall McLuhan, *Understanding Media: The Extensions of Man* (Cambridge, MA: The MIT Press, 1994).

30 Barry Commoner, *The Closing Circle: Nature, Man, and Technology* (New York: Random House, 1971).

31 Guien, *Obsolescences*, p. 306.

NOTES TO PP. 226–232

32 Donella H. Meadows, Dennis L. Meadows, Jørgen Randers and William W. Behrens, *Limits to Growth* (New York: Potomac Associates, 1972).

33 Rather than submerging our readers in an endless list of works that question sustainable development, we will simply refer to Fanny Verrax's recent article on light bulbs (our old friends!), in which she shows how the issues of energy efficiency, recyclability and lifespan have found concrete expression: Fanny Verrax, 'Illuminating the light bulb', *Techniques & Culture*, no. 65–6, 2016, available online at: http://journals.openedition.org/tc/7842.

34 https://themaintainers.org/maintainers-ii-2017.

35 Technical page 'Économie circulaire. Notions', p. 7, available online at: https://www.ademe.fr.

36 https://www.europarl.europa.eu/doceo/document/A-8-2017-0214_FR.html.

37 https://repair.eu/news/analysis-of-the-adopted-directive-on-common-rules -promoting-the-repair-of-goods/.

38 The business models that are in the process of being invented will deserve, in this regard, close study if we are to understand how certain forms of durability and reparability in consumer goods are valued and concretely produced in factories and on supply chains.

39 Nicky Gregson, Alan Metcalfe and Louise Crewe, 'Practices of object maintenance and repair: How consumers attend to consumer objects within the home', *Journal of Consumer Culture*, vol. 9, no. 2, 2009, pp. 248–72.

40 https://www.ifixit.com/Manifesto.

41 Grinvald and Tur-Sinai, 'The right to repair'.

42 Kyle Wiens, 'New high-tech farm equipment is a nightmare for farmers', *Wired*, 2015, available online at: https://www.wired.com/2015/02/new-high-tech -farm-equipment-nightmare-farmers/#:~:text=%E2%80%9CThere's%20an %20increasing%20number%20of,require%20a%20computer%20to%20fix.%E2 %80%9D&text=The%20problem%20is%20that%20farmers,the%20keys%20to %20those%20boxes.

43 For an overview of open-source initiatives in agriculture, see Quentin Chance and Morgan Meyer, 'L'agriculture libre. Les outils agricoles à l'épreuve de l'open source', *Techniques & Culture*, no. 67, 2017, available online at: http://journals .openedition.org/tc/8511.

44 Wiens, 'New high-tech farm equipment'.

45 The terms of this struggle resonate directly with the reflections of Gilbert Simondon on technicism in general: its twofold mechanism of enslavement, of both nature and humans, means the latter find themselves 'alienated' in an asymmetrical relationship to technical objects. See Gilbert Simondon, *On the Mode of Existence of Technical Objects*, translated by

Cecile Malaspina and John Rogove (Minneapolis, MN: University of Minnesota Press, 2016).

46 Matthew B. Crawford, *Shop Class as Soulcraft: An Inquiry into the Value of Work* (London: Penguin, 2009); Richard Sennett, *The Craftsman* (London: Penguin, 2009).

47 Daniela K. Rosner and Fred Turner, 'Theaters of alternative industry. Hobbyist: Repair collectives and the legacy of the 1960s American counterculture', in Hasso Plattner, Christoph Meinel and Larry Leifer (eds.), *Design Thinking Research: Understanding Innovation* (New York: Springer, 2015), pp. 59–69.

48 Steward Brand, *How Buildings Learn: What Happens after They're Built* (New York: Viking Penguin, 1994).

49 Grinvald and Tur-Sinai, 'The right to repair'; Hatta, 'The right to repair'. For a detailed description of the practices of diagnosis and the forms of resistance of independent garage owners in France, see Mallard, 'Être garagiste', and Dutertre and Jullien, 'Les artisans de la réparation automobile'.

50 https://www.congress.gov/bill/112th-congress/house-bill/1449.

51 Lara Houston, 'Unsettled repair tools: The "death" of the J.A.F. box', *The Maintainers: A Conference*, 2016, available online at: http://themaintainers.wpengine.com/wp-content/uploads/2021/04/Maintainers-Lara-Houston.pdf.

52 Nicolas Nova and Anaïs Bloch, *Dr. Smartphone: An Ethnography of Mobile Phone Repair Shops* (Geneva: Idpure, 2020).

53 Alain Tarrius, *La Mondialisation par le bas. Les nouveaux nomades de l'économie souterraine* (Paris: Balland, 2002).

54 Nova and Bloch, *Dr. Smartphone*, p. 182.

55 Albena Yaneva, 'How buildings "surprise": The renovation of the Alte Aula in Vienna', *Science Studies*, vol. 21, no. 1, 2001, pp. 8–28 (p. 17).

56 Ademe, *Les Français et la réparation. Perceptions et pratiques* – 2019 edition: Final Report 2020.

57 Kyle Wiens, 'The right to repair will help us endure outbreaks', *Wired*, 2020, available online at: https://www.wired.com/story/opinion-the-right-to-repair-will-help-us-endure-outbreaks/.

58 See chapter 2.

59 Madeleine Akrich, 'The de-scription of technical objects', in Wiebe E. Bijker and John Law (eds.), *Shaping Technology/Building Society Studies in Sociotechnical Change* (Cambridge, MA: The MIT Press, 1992), pp. 205–24.

60 Steven J. Jackson, 'Rethinking repair', in Tarleton Gillespie, Pablo J. Boczkowski and Kirsten A. Foot (eds.), *Media Technologies: Essays on Communication, Materiality, and Society* (Cambridge, MA: The MIT Press, 2014), pp. 221–40.

61 Nova and Bloch, *Dr. Smartphone*, pp. 185–6.

NOTES TO PP. 238–245

62 Anna Lowenhaupt Tsing, *The Mushroom at the End of the World: On the Possibility of Life in Capitalist Ruins* (Princeton, NJ: Princeton University Press, 2015).

63 Donna Haraway, *The Companion Species Manifesto: Dogs, People, and Significant Otherness* (Chicago, IL: University of Chicago Press, 2003).

64 Aryn Martin, Natasha Myers and Ana Viseu, 'The politics of care in techno-science', *Social Studies of Science*, vol. 45, no. 5, 2015, pp. 625–41.

65 See chapter 5.

66 Lazar Kunstmann, *Urban eXperiment* (Paris: Uqbar, 2018). (Jacques-Germain Soufflot was the architect who designed the Panthéon.)

67 Ibid., p. 41.

68 Gilles Thomas, 'La fa(r)ce cachée des Grandes Écoles. Les "catacombes" offertes à leurs élèves', *In Situ*, no. 17, 2011, available online at: https://journals.openedition.org/insitu/1213.

69 Caroline Ibos, 'Mierle Laderman Ukeles et l'art comme laboratoire du *care*. "Lundi matin, après la révolution qui s'occupera des poubelles?"', *Cahiers du Genre*, vol. 66, no. 1, 2019, pp. 157–79 (p. 165).

Conclusion

1 Bernadette Bensaude-Vincent, *Temps-paysage. Pour une écologie des crises* (Paris: Le Pommier, 2021).

2 On the importance, for a pluralist composition of the world, of a diplomacy that reactivates the art of immanent attention, see Isabelle Stengers, 'The challenge of ontological politics', in Marisol de la Cadena and Mario Blaser (eds.), *A World of Many Worlds* (Durham, NC: Duke University Press, 2018), pp. 83–111.

3 Philippe Descola, 'La forêt des signes', in Eduardo Kohn, *Comment pensent les forêts. Vers une anthropologie au-delà de l'humain* (Paris: Zones sensibles, 2017), pp. 11–17.

4 Our thanks to Jean Danielou, who invited us to explore this semiotic path by bringing us back into the conversation between the two anthropologists.

5 Bruno Latour, *La Clef de Berlin. Petites leçons de sociologie des sciences* (Paris: La Découverte, 1993).

6 Pascale Molinier, Sandra Laugier and Patricia Paperman (eds.), *Qu'est-ce que le care? Souci des autres, sensibilité, responsabilité* (Paris: Payot, 2009); Joan C. Tronto, *Moral Boundaries: A Political Argument for an Ethic of Care* (New York: Routledge, 1993).

7 Ruth Schwartz Cowan, *More Work for Mother: The Ironies of Household Technology from the Open Hearth to the Microwave* (New York: Basic Books,

1983). Furthermore, the material care provided by women goes far beyond the domestic sphere, as shown in particular by Delphine Gardey, *Le Linge du Palais-Bourbon. Corps, matérialité et genre du politique à l'ère démocratique* (Lormont: Le Bord de l'eau, 2015).

8 We must insist on the circumscribed nature of the itinerary that we have proposed, which points to many sites for readers to explore or revisit. We have not covered all maintenance situations and practices – far from it. In particular, we have barely mentioned maintenance activities in industry, or those practised in the poorest regions of the world. The relative exclusion of these two configurations, which might be thought central to the questioning of maintenance, has allowed us to follow in the footsteps of the 'maintenance art' of Mierle Laderman Ukeles and to underline the political dimensions which run through the question of the duration of things beyond, and in parallel with, very specific issues of risk (in industry) and material and economic constraints (in the Global South).

9 Tronto, *Moral Boundaries*.

10 Caroline Ibos, 'Mierle Laderman Ukeles et l'art comme laboratoire du *care*. "Lundi matin, après la révolution qui s'occupera des poubelles?"', *Cahiers du Genre*, vol. 66, no. 1, 2019, pp. 157–79 (p. 171).

11 Emmanuel Bonnet, Diego Landivar and Alexandre Monnin, *Héritage et fermeture. Une écologie du démantèlement* (Paris: Divergences, 2021).

12 These, first and foremost nuclear installations and their waste, form the most delicate part of what Bonnet and his colleagues call the 'negative commons', for which they emphasize the importance of inventing new forms of inheritance.

13 Catherine Larrère, 'Environmental ethics: Respect and responsibility', in Aurélie Choné, Isabelle Hajek and Philippe Hamman (eds.), *Rethinking Nature: Challenging Disciplinary Boundaries* (Abingdon and New York: Routledge, 2017), pp. 54–83.

14 John Baird Callicott, *In Defense of the Land Ethic: Essays in Environmental Philosophy* (Albany, NY: State University of New York Press, 1989), p. 110.

15 Larrère, 'Environmental ethics'.

16 Donna J. Haraway, 'Modest_ witness@second_millennium', the first chapter in her *Second Millennium.FemaleMan© Meets OncoMouseTM. Feminism and Technoscience* (New York and London: Routledge, 1997). See also Maria Puig de la Bellacasa, '"Nothing comes without its world": Thinking with care', *The Sociological Review*, vol. 60, no. 2, 2012, pp. 197–216.

17 John Dewey, *Theory of Valuation* (Chicago, IL: University of Chicago Press, 1939). In an article that brings together their different studies, Lara Houston and her colleagues propose analysing certain repair situations as 'sites of valuation' where distinct values are expressed and sometimes come into conflict. See Lara

NOTES TO PP. 248–249

Houston et al., 'Values in repair', *CHI'16 Proceedings of the 2016 CHI Conference on Human Factors in Computing Systems* (New York: ACM, 2016), pp. 1403–14.

18 William Cronon, 'The trouble with wilderness, or getting back to the wrong nature', in William Cronon (ed.), *Uncommon Ground: Rethinking the Human Place in Nature* (New York: W. W. Norton, 1995), pp. 69–90.

19 Ibid., p. 86.

20 Ibid., p. 89.

Index

abandonment, 239–40
acceleration, 30
   Covid-19 pandemic, 32
accidents *see* breakdowns
action, delegation of, 49, 51, 52–3
'Adventure of the Cardboard Box, The'
   (Conan Doyle), 77–8
advertising *see* marketing
aeroplane crashes, 23
aesthetics, 56
agriculture, 204–5, 216
see also John Deere tractors
amateurs, 89
Anders, Günther, 239
Anthropocene, the, 21
anthropology, 50
anxiety, 109, 170, 172, 207–8
Apple, 215–16, 217
archaeology, 142–3, 188
architecture, 31, 104
   see also building renovation/
     restoration
art, 16–17
   attributing, 77
   authenticity, 177
   conservation, 150–4, 155–6
   exhibiting, 152–3, 173
   freeports, 156
   restoration, 156, 157
attachment, 86–90, 110, 165–6, 177–8
attention, 5–6, 7–9, 13, 92–3
   attachment, 86–90, 110
   displacements, 66–71
   expertise, 76–83
   fragility, 62, 70, 71

inspection rounds, 62–6
multisensoriality, 71–6, 89–90
reassignment of, 37–40
vigilance, 83–6
see also inspection rounds
Austrian National Library, 147–9
authenticity, 169–71
   adjustments, 171–9
   anxiety, 170, 172, 207–8
   attachment, 177–8
   conservation, 189–97
   environmental ethics, 198–207, 210
   ethics, 207–10
   heritage diplomacy, 184–98
   minimal intervention, 196–8
   options, 175–6
   resistance, 181, 182
   restoration, 191–6
   reversibility, 176
   surprises, 179–84
autonomy, 60, 61

battery charging, 3
Bensaude-Vincent, Bernadette, 166
Bessy, Christian and Chateauraynaud,
   Francis
   Experts et faussaires (*Experts and
     Counterfeiters*), 169
bicycles, 38–9
   Covid-19 pandemic, 39
   maintenance, 3, 39–40
   repair, 39
Bidet, Alexandra
   L'Engagement dans le travail
     (*Commitment at Work*), 69

INDEX

biocentric logic, 247

blackouts, 24

bodies
of maintainers, 94, 96–7
preservation of, 131–2, 133–4, 136–40

Boito, Camillo
Conservare o restaurare (*Conserve or Restore*), 192

Bonnet, Emmanuel, Landivar, Diego and Monnin, Alexandre, 6, 246–7

books, 148–9

Brandi, Cesare, 195–6
Teoria del restauro (*The Theory of Restoration*), 196

breakdowns, 2, 21–7
'broken world thinking', 28–9
electricity, 24–5
infrastructure monitoring, 66–9
Jackson, Steve, 28
Oliver, John, 26, 34
revelations, 23–5

'broken world thinking', 28–9

building renovation/restoration, 2, 46–7, 181–3, 184–90

Callén, Blanca and Sánchez Criado, Tomás, 100

Callicott, John Baird, 201–2, 203

care, 58–61, 110

cars, 125
authenticity, 171–7, 178–81, 183, 187
Chevrolet, 223
disassembly, 98
emotions, 108–9
environmental issues, 176–7
Ford Mustang, 89–90, 171–81, 183, 187
mass consumption, 222–4
mechanics, 75–6, 91, 98, 108–9
Model T Ford, 215, 222–3
OBD-II emissions diagnostic tool, 234
right to repair, 215, 234

Challenger shuttle disaster, 22

Chartres Cathedral, 184–9, 193

Chateauraynaud, Francis, 85, 86

Chevrolet, 223

choreography *see* dance metaphor

circular economy, 7, 226–7

civilization, 202, 225

cleaning, 104–5

Clément, Gilles, 205–6

clothing, mending, 3

clue paradigm, 77–8, 79, 80

coexistence, ethics of, 208–9

cognition, 50

Cohn, Marisa, 160–5

Commoner, Barry, 225–6

Conan Doyle, Arthur, 77
'Adventure of the Cardboard Box, The', 77–8

conflicts, 15, 211
duration, values of, 224–7
employment, 236–7
humans, 236–8
knowledge, redistributed, 233–6
medical equipment, 211–15
planned obsolescence, 218–27
responsibility, 238–41
right to repair, 215–16, 217–18, 226, 227–38, 240–1
use, emancipation of, 227–33

connoisseurs, 79, 86–7

Conservare o restaurare (*Conserve or Restore*) (Boito), 192

conservation, 115, 155, 189–97
art, 150–4, 155–6
authenticity, 189–97
books, 148–9
Chartres Cathedral, 184–9, 193
DeSilvey, Caitlin, 158–60, 164
environmental, 199
heritage, 189–97
post-preservationist, 158–60, 164
scraping, 192, 196

consumerism, 7, 88, 221–2, 232
see also consumption

INDEX

consumption, 7, 88, 170, 219, 220–6, 228, 232, 242

copyright, 230

Covid-19 pandemic, 31–3, 39, 211–14, 215, 237

Crawford, Matthew, 94–5
Shop Class as Soulcraft, 25, 94–5

crises, 5, 6, 21, 57

Cronon, William, 248

Curated Decay (DeSilvey), 158–60

dance metaphor, 110–13, 137–8, 144, 183–4, 208, 209

Dant, Tim, 75, 91–2, 109

Dark Abyss of Time, The (Olivier), 141–3

de Laet, Marianne and Mol, Annemarie, 103

defects, 43–4

delegation, 49, 51

descaling, 1

Descola, Philippe, 243–4

DeSilvey, Caitlin, 158, 164
Curated Decay, 158–60

Despret, Vinciane, 261 n. 26

deterioration, 56–7, 151–5, 160
books, 148, 149
DeSilvey, Caitlin, 158, 159, 160
Lenin, Vladimir, remains of, 131, 132
Mona Lisa, 150–1, 152–4
Paris metro signage, 43–5, 132, 134, 135
slowdown, 147–55, 156
Wagner clock, 116, 120, 156

digital companies, 24

Digital Millennium Copyright Act (DMCA), 230, 231

digitalization, 229

diplomacy, 243
see also heritage diplomacy

disaggregation, 100

disassembly, 98–102

disconnectionist ecology, 246

disposable products, 220–1

disruption, 18, 45–6

DMCA (Digital Millennium Copyright Act), 230, 231

domestic work, 37, 245

Domínguez Rubio, Fernando, 150, 151, 152

Dr. Smartphone (Nova and Bloch), 98

duration, 224–7

Ecce Homo (García Martínez, Elías), 157

ecofeminism, 210

ecology, 58

Edensor, Tim, 46–7

Edgerton, David, 19–20
Shock of the Old, The, 19

Eiffel Tower, 137–8

electricity
breakdowns, 24–5
light bulbs, 219–20
networks, 2

electronic device repair, 3, 98, 100–2, 109

emotions, 108–9, 164, 170, 172, 207–8

employment, 236–7

encounters, 13–14

Ending the Depression through Planned Obsolescence (London), 224

L'Engagement dans le travail (*Commitment at Work*) (Bidet), 69

environment, the, 5, 6, 159–60, 198–207, 246–9
biocentric logic, 247
cars, 176–7
circular economy, 7, 226–7
crisis, 5, 6, 21, 57
ethics, 198–207, 210
human relationship with, 6, 201–6
obsolescence, 225–6
recycling, 96, 100–2
sensitivity to, 5, 6
sustainability, 204–5, 226
technology, relying on, 25
see also waste management

285

INDEX

ethics, 60, 207–10
  of coexistence, 208–9
  environmental, 198–207, 210
Europe, 216, 227
expertise, 76–83
Experts et faussaires (*Experts and Counterfeiters*) (Bessy and Chateauraynaud), 169

'Farmer as a Conservationist, The' (Leopold), 204
FDA (Food and Drug Administration), 214
fire alarms, 2
Fisher, Berenice and Tronto, Joan, 60
flamingos, 206
fluid technology, 103
Food and Drug Administration (FDA), 214
Ford, Henry, 215, 222–3
Ford Mustang, 89–90, 171–81, 183, 187
forestry, 206
fragility, 13, 46–8, 52, 53–4, 80–1
  attention, 62, 70, 71, 82, 92, 93
  care, 58–61
  consumerism, 88
  of maintainers, 97
  materials, 46, 47–8, 54, 58
  Paris metro signage, 45
  perception of, 70–1, 87–8
  Saint Ann's Church, Manchester, 46–7
  template documents, 82
  wear and tear, 55–7, 58, 62
France, 141–3, 218–19, 226–7
  see also Paris
freeports, 156
Fürst, Moritz, 147–9

Gagarin (Liatard and Trouilh), 87–8
gender, 10
Geymüller, Heinrich von, 195
Gibson, James J., 70

Gillette, 220, 221
Ginzburg, Carlo, 77, 79–80
graffiti removal, 73–5, 79, 105, 106–8
Gregson, Nicky, 127

Halliday, Johnny, 181
Hartog, François, 123–4
health, 59–60, 211–15
Heidegger, Martin, 11–12, 22–3
Henk, Chris and Sims, Ben
  Repairing Infrastructures, 22
heritage, neglect of, 239–41
heritage conservation, 189–97
  authenticity, 189–97
  Chartres Cathedral, 193
  environmental, 199
  scraping, 192, 196
heritage diplomacy, 184–98
  Boito, Camillo, 192
  Brandi, Cesare, 195–6
  Chartres Cathedral, 184–9, 193
  Geymüller, Heinrich von, 195
  minimal intervention, 196–8
  Viollet-le-Duc, Eugène, 190, 191–3, 194
heroism, 34–5, 37
Hidden Persuaders, The (Packard), 225
historicity, 123–4
history, 122, 143
horticulture, 205–6
hospitals, 211–15
Houston, Lara, 98, 102, 109, 234–5
How Forests Think (Kohn, Eduardo), 243–4
humans, 236–8
  environment, relationship with, 6, 201–6
  intervention of, 4, 158–60, 171, 194–5, 198–9, 202–6, 208
  objects, relationship with, 28, 51, 58
  things, relationships with, 5–7, 8, 243
  time, relationship with, 192
  vocabulary, 10

286

INDEX

Hummel, Cornelia, 89–90, 171–3, 177, 179, 187
Hutchins, Edwin, 50

I Make Maintenance Art One Hour Every Day (Ukeles), 35–6
IBM, 215
iFixit, 212, 214, 215
Ingold, Tim, 12, 54, 111–12, 209
innovation, 17–21, 22, 38
Innovation Delusion, The (Russell and Vinsel), 18
inspection rounds, 62–6
  graffiti removal, 73–5, 79
  libraries, 148–9
  Mona Lisa, 154
  monitoring instruments, 80–2
  movement, 69–71
  Paris metro, 62–3, 65, 70–1, 78, 79, 81, 135
  pest control, 62, 64–5, 78, 79, 80, 81–2
  photocopiers, 62, 63–4, 70, 71–2, 78, 79, 81
  proximity, 66–70, 72, 80–1
  Wagner clock, 116
  water networks, 62, 63, 72–3, 78, 79, 81, 84–5
intellectual property, 229–31
interdependencies, 24–5
intuition, 80
Ise Shinto temples, 144–6, 197
IT sector, 215

Jackson, Steve, 28–9, 30, 31
  'Rethinking Repair', 28
James, Henry
  Little Tour in France, A, 192
Jeannot, Bernard, 114–15, 118
John Deere tractors, 216, 230–1, 232

kites, 111–12
knowledge
  models, 77, 87

right to, 233–6
Kohn, Eduardo
  How Forests Think, 243–4
Krajewski, Markus, 219
Kunstmann, Lazar, 115, 116, 117, 120, 122, 239–40

Last Week Tonight (Oliver), 26
Latour, Bruno, 6–7, 11, 51
  'Parliament of Things', 57
  'sociologie sans objet? Remarques sur l'interobjectivité, Une' ('A sociology without an object? Remarks on interobjectivity'), 49
  Star, Susan Leigh, 52–3
Lenin, Vladimir, remains of, 130–2, 133–4, 136–7, 138, 139–41
Leonardo da Vinci, 151
  Mona Lisa, 150–4
Leopold, Aldo, 201, 204, 207
  'Farmer as a Conservationist, The', 204
  'Popular Wilderness Fallacy, The', 201
libraries, 148–9
lifespans, 128–9, 216, 222–3, 224–7
  see also planned obsolescence; prolongation
light bulbs, 219–20
Little Tour in France, A (James), 192
London, Bernard
  Ending the Depression through Planned Obsolescence, 224

maintainers, 26, 35–7, 54, 209–10, 244–6
  activists, 239
  amateur, 89–90
  bodies of, 94, 96–7
  as connoisseurs, 79, 86–7
  details, attention to, 78–9
  domestic, 245
  emotions, 108–9, 164

INDEX

maintainers (*cont.*)
  employment, 236–7
  expertise, 76–83
  matter, 55
  mechanics, 75–6, 91, 98, 108–9
  monitoring instruments, 80–2
  multisensoriality, 71–6, 89–90, 91
  Paris metro signage, 42–4, 45, 99
  restorers, 46–7
  Star, Susan Leigh, 53
  suffering, 95–7
  template documents, 82, 83
  vigilance, 83–6
  see also repairers
Management Guidelines for World
  Cultural Heritage Sites (Iccrom
  and Icomos), 196
marketing, 222–3, 225
material(s), 12, 107–8
  composing with, 178–9
  fragility, 46, 47–8, 54, 58
  Ingold, Tim, 54
  oil paintings, 150, 152
  Paris metro signage, 41–2, 44
  stone, 46–7
  wear and tear, 55–7
materiality, 54
Mathevet, Raphaël and Béchet,
  Arnaud, 206
matter, 54, 55
Matters of Care (Puig de la Bellacasa),
  73
mechanics, 75–6, 91, 98, 108–9
medical equipment, 211–15
Medical Imaging and Technology
  Alliance, 214
medicine, 59–60
mending, 126, 127
Mérimée, Prosper, 190
Mexicaine de Perforation (Mexican
  Consolidated Drilling Authority),
  115, 116, 119
militant actions, 128, 129

minimal intervention, 196–8
Minneapolis bridge collapse, 22
Model T Ford, 215, 222–3
modernity, breakdown of, 21
Moffit, David, 67–8
Mol, Annemarie, 59
Mona Lisa (Leonardo da Vinci), 150–4
monitoring instruments, 80–2
Monnet, Pascal, 118, 119, 120–1, 122
Morelli, Giovanni, 77
Morizot, Baptiste, 5, 57
  disruption, 45–6
  Ways of Being Alive, 5
Morris, William, 191
Muir, John, 200–1
multisensoriality, 71–6, 89–90
Musée du Quai Branly, 64
museums, 150–3, 173

Nara Document on Authenticity, 197
nature, 199, 200–7, 210, 243–4
New York City, 17
  Department of Sanitation, 36–7
Norman, Don, 50
Nova, Nicolas and Bloch, Anaïs, 235
  Dr. Smartphone, 98
nuclear waste, 166–7, 281 n. 12

objects, 10–12, 233
  agency of, 51, 52–3
  anthropology, 50
  delegation, 49, 51, 52–3
  interdependencies, 24–5
  internal components, 23–4
  politics, 51–2, 53
  'presence to hand', 22–3
  psychology, 50
  'readiness to hand', 22–3
  relationships with, 28, 51, 58
  social sciences, 48–55
  solidity, 52, 53–4, 55, 58
  transformations, 102–6
  types, 18–19

INDEX

worn-out, 19–20
see also things
obsolescence, 19
  planned, 218–27
Oliver, John, 26–7, 34, 35
  Last Week Tonight, 26
  proximity, 67
Olivier, Laurent
  Dark Abyss of Time, The, 141–3
ontology, 124, 170–1
  permanence, 124, 129–47
  prolongation, 124, 125–9, 161–4
  slowdown, 124–5, 147–57
  stubbornness, 125, 157–67
Oradour-sur-Glane, 141–3
ownership, 228–31

Packard, Vance, 225
  Hidden Persuaders, The, 225
  Waste Makers, The, 225
Panthéon, 114–15, 116–21, 122–3, 156,
  168, 239–40
Paris
  Eiffel Tower, 137–8
  graffiti removal, 73–4, 79, 105, 106–8
  metro inspection rounds, 62–3, 65,
    70–1, 78, 79, 81, 135
  metro signage, 41–5, 99, 132–6, 138–9
  Panthéon, 114–15, 116–21, 122–3, 156,
    168, 239–40
'Parliament of Things', 57
perception, 70–1
perfection, 56
permanence, 124, 129–47
  dance metaphor, 137–8
  Ise Shinto temples, 144–6
  Lenin, Vladimir, remains of, 130–2,
    133–4, 136–41
  Oradour-sur-Glane, 141–3
  Paris metro signage, 132–6, 138–9
  pretence of, 143–4
  work involved in, 135–7
pest control, 62, 64–5, 78–82

Philips, Anton, 220
philosophy, 54–5
Phoebus cartel, 219–20
phone repair, 3, 98, 102, 109, 235
photocopiers, 62, 63–4, 70, 71–2, 78,
  79, 81
Pickering, Andrew, 111–12
Pinchot, Gifford, 200–1
planned obsolescence, 218–27
plumbing, 1–2, 11
politics, 128, 129, 242–8
  objects, 51–2, 53
pollution, 225
'Popular Wilderness Fallacy, The'
  (Leopold), 201
posthumanism, 54–5
preservation
  of bodies, 131–2, 133–4, 136–40
  of nature, 199, 200–1, 202–4, 206
  Oradour-sur-Glane, 141–3
  pretence of permanence, 143–4
  Sengu tradition, 144–6
prestige, lack of, 17
pretension, 129
priorities, 247
prolongation, 124, 125–9, 161–4,
  217–18
proximity, 66–70, 72, 80–1
psychology, 50
Puig de la Bellacasa, Maria
  Matters of Care, 73

railways, 2
razors, 220, 221
reality, 47–8
recalcitrance, 94–7
recycling, 96, 100–2
relationships, 5–6
  environment, 6, 201–6
  objects, 28, 51, 58
  things, 5–7, 8, 243
  time, 192
repair, 20–6, 27–8, 33, 34, 38

289

INDEX

repair (*cont.*)
  activists, 232–3
  bicycles, 39
  'broken world thinking', 28–9
  electronic devices, 3, 98, 100–2, 109
  etymology, 27
  Jackson, Steve, 28–9
  knowledge, right to, 233–6
  manuals, 212–15
  medical equipment, 211–15
  Oliver, John, 26
  phones 3, 98, 102, 109, 235
  right to repair, 215–16, 217–18, 226, 227–38, 240–1
  struggle, the 34
  see also repairers
Repair Cafés, 232–3
repairers, 34
  employment, 236–7
  heroism, 34–5
Repairing Infrastructures (Henk and Sims), 22
repetition, 30, 33
resilience, 27
resistance, 92–4
  authenticity, 181, 182
  dance metaphor, 110–13
  disassembly, 98–102
  recalcitrance, 94–7
  transformations, 102–6, 109–10
  worries, 106–10
responsibility, 238–41
restoration, 120, 121–2, 156–7
  art, 156, 157
  authenticity, 191–6
  building, 2, 46–7, 181–3, 184–90
  clandestine, 115–19, 120–2, 156, 239
  of time, 121–2
restrictions, 212, 213, 214
'Rethinking Repair' (Jackson), 28
return, 27, 28, 29, 33
revelations, 23–5
Rifat, Mohammad Rashidujjaman, 96

right to repair, 215–16, 217–18, 226, 227–38, 240–1
Rotor collective, 55–8
  Usus/usures. États des lieux/How Things Stand, 56
routine, 26–34, 125–6
Ruskin, John, 191
Russell, Andrew and Vinsel, Lee, 18, 22
  Innovation Delusion, The, 18

sabotage, 114, 119–21
Saint Ann's Church, Manchester, 46–7
Sample, Hilary, 31
Saraç-Lesavre, Başak, 166
Savalli, Anne, 184–9, 193
science, 111
scraping, 192, 196
semiotics, 243–4
Sengu tradition, 144–5
senses, the, 71–6, 89–90, 91, 107
Shaw, Rob, 104
Shock of the Old, The (Edgerton), 19
Shop Class as Soulcraft (Crawford), 25, 94–5
Sloan, Alfred, 222, 223
slowdown, 124–5, 147–57
social sciences, 48–55
'sociologie sans objet? Remarques sur l'interobjectivité, Une' ('A sociology without an object? Remarks on interobjectivity') (Latour), 49
software, 229–31
spacecraft, 160–5
  Challenger shuttle disaster, 22
Spain, 100
spare parts, 101–2, 109
Spelman, Elisabeth, 34
standardization, 41, 132
Star, Susan Leigh, 52–3
Statue of Liberty, 67–8, 69
stone, 46–7
stories, 8–9

INDEX

street cleaning, 104–5
see also graffiti removal
stubbornness, 125, 157–67
sustainability, 204–5, 226
Switzerland, 235

tact, 14–15, 208–9
technology, 19–20
breakdowns, 22, 23, 24–5
Covid-19 pandemic 31–3, 211–12
digitalization, 229
electronic device repair, 3, 98, 100–2, 109
fluid, 103
interdependencies, 24–5
IT sector, 215
phone repair, 3, 98, 102, 109, 235
'presence to hand', 22–3
progress, 223
prolongation, 128
'readiness to hand', 22–3
repair activists, 232
right to repair, 215–16, 217, 229
software, 229–31
space, 160–5
time, 30–2
temporality *see* time
Teoria del restauro (*The Theory of Restoration*) (Brandi), 196
things, 6–7, 10–12, 233
living, 57
relationships with, 5–7, 8, 243
sensitivity to, 6
stubbornness of, 165
vocabulary, 10–11
see also objects
Thoreau, Henry, 200
time, 14, 30–3, 123, 124, 167–8
Hartog, François, 123–4
permanence, 124, 129–47
prolongation, 124, 125–9, 161–4, 217–18
relationship with, 192

restoration, 121–2
slowdown, 124–5, 147–57
stubbornness, 125, 157–67
Wagner clock, 114–15, 116–21, 122–3, 168
Touch Sanitation (Ukeles), 36–7
Transfer: The Maintenance of the Art Object: Mummy Maintenance: With the Maintenance Man, the Maintenance Artist, and the Museum Conservator (Ukeles), 16–17
transformations, 102–6, 109–10
travellers, comfort of, 133
Tronto, Joan, 60, 61, 245

Uganda, 98, 102, 109, 234–5
Ukeles, Mierle Laderman, 15, 29, 35, 37
dance metaphor, 111
I Make Maintenance Art One Hour Every Day, 35–6
responsibility, 241
Touch Sanitation, 36–7
Transfer: The Maintenance of the Art Object: Mummy Maintenance: With the Maintenance Man, the Maintenance Artist, and the Museum Conservator, 16–17
Washing/Tracks/Maintenance: Outside 29–30
United States, 199–200, 210, 213, 214, 216–17
University of Vienna, 181–3, 184
Untergunther, 115, 117, 120, 122, 156, 239–41
Urban eXperiment (UX) *see* UX
urban planning, 17, 38, 104–5
graffiti removal, 73–5, 79, 105, 106–8
US Public Interest Research Group, 213
use, 228
Usus/usures. États des lieux/How Things Stand (Rotor collective), 56

INDEX

UX (Urban eXperiment), 115–19, 122
  see also Untergunther

valuation, 248
Venice Charter, 196
vigilance, 83–6
violence, 95, 96, 97
Viollet-le-Duc, Eugène, 190, 191–3,
  194
Viot, Jean-Baptiste, 116–20, 123
vocabulary, 10–12
vulnerability tests, 100

Wagner clock, 114–15, 116–21, 122–3,
  156, 168, 239–40
Waldman, Jonathan, 67
Washing/Tracks/Maintenance: Outside
  (Ukeles), 29–30

Waste Makers, The (Packard), 225
waste management, 96, 100–2, 104–5,
  225–6
  nuclear, 166–7, 281 n. 12
water networks, 62, 63, 72–3, 78, 79,
  81, 84–5
water pumps, 102
Ways of Being Alive (Morizot), 5
wear and tear, 55–7, 58, 62
Wiens, Kyle, 212, 214, 215, 226, 237
wilderness, 199–202, 203, 248
women, 210, 245
writing, power of, 41–2

Yaneva, Albena, 181–3, 184
Yurchak, Alexei, 130

Zimbabwe, 103